FAILING FORWARD FAST

What 25 Years in the CIA Taught
Us about Getting Things Done in
Bureaucracies

Bruce Hartmann
and
Greg Moore

Failing Forward Fast

KJI Ltd
PO Box 9257
Pueblo, CO 81008

Printed in the United States of America

ISBN-13: 978-1515264248
ISBN-10: 1515264246

To Joe, who hired us, and to Steve, who introduced us to failing
forward fast.

Throughout this book we use the terms "we" and "us" to mean either one of us or both of us for any given instance.

CONTENTS

Preface:
Getting the Most out of Your Time in Government vii

Introduction:
Failing Forward Fast...9

Chapter One
Working at the Central Intelligence Agency......................21

Chapter Two.
Creating a More Effective Work Environment 27

Chapter Three
Personal Skills to Put You Ahead of the Crowd................45

Chapter Four
Self-Deception...64

Chapter Five
Failing-Forward-Fast Program Management....................77

Chapter Six
Contracts, Legal, and Unwritten Rules...........................97

Chapter Seven
Some Things You Should Never Do...............................110

Chapter Eight
Working Your Way out of a Mess................................121

Chapter Nine
Staying Healthy in a Tough Job.......................136

Chapter Ten
Some Thoughts about Retirement......................142

Parting Comments
A Memorandum for the Future Director of CIA..............149

Getting the Most out of Your Time in Government

This book is focused mainly on readers who were educated as scientists and engineers, and later became government employees who now manage technical work done through contractors. Our intent is to suggest some ways in which you can work better, think more effectively, and create a more innovative environment that draws the best people to your problems. We wish we had a copy when we started our government careers.

To be clear, however, we are not trying to provide you with a recipe book. We are simply going to tell you what worked for us as technologists for over 25 years in the Central Intelligence Agency, beginning before the fall of the Berlin Wall and ending just before the Abbottabad Raid that led to the death of Bin Laden. Our main goal is to help you get the most out of your time in government. Take the lessons we learned and adapt what makes sense for your environment.

Failing Forward Fast

Failing forward fast is a bit like selectively kissing frogs until you find the prince.

Both of us were in our mid-30s when we decided to join the Agency. One of us was an electrical engineer and the other a materials scientist, and each of us had over ten years of industrial and academic experience. We still think that the mid-30s is the best age to join government. You'll be more effective if you first get some experience on the outside, build a few things, and then spend some of your remaining time giving back to the country that allowed you to do all of that.

The CIA was the only government organization we really considered working for. Perhaps it was too many movies or maybe it was just curiosity. In the end, however, there was no better way to find the ground truth than to join the organization.

We were hired to be Contracting Officer Technical Representatives (COTRs) in the Office of Research and Development, or ORD. While that sounds very boring and bureaucratic, it was anything but that. Being a COTR was a lot like running your own business, except it was done within the government rather than for industry. Agency COTRs identified customers, determined needs, found money, found contractors, worked with lawyers, engaged contracting officers, managed contracts, accepted and tested the resulting products, and sometimes helped with deployment in operations.

Being in ORD also allowed us to continue to work in research and development rather than, say, becoming an analyst and losing some of that hands-on exploration that we loved. It also involved us in operations which gave us invaluable insight into that world. During our first year with the Agency, for example, we were assigned to an operation that involved rescuing someone who had been kidnapped by a terrorist group. The specific problem had not been faced by the operations group before, and required the quick development of some new technology. No one had ever asked us to help save someone's life, and it was then that we decided this was the place to make a career. The opportunities to make important contributions seemed endless.

We showed up near the end of an era for the Agency. Bill Casey was the boss - the Director of Central Intelligence (DCI) - and the last of the old guard that had been around when the Agency was the Office of Strategic Services. Even though we were many, many layers below Casey's radar, we enjoyed working for the guy. You could feel his influence all the way down to the bottom. That would not be the case for many of the Directors later in our careers. Nonetheless, we managed to get a few good things done in our 25-year careers with the Agency. We'll tell you about some of them in this book, and explain what it took to get them done in the CIA's bureaucracy.

Why We Wrote This Book

When we first joined the CIA, we were sent to many courses,

which was a good thing because we really didn't have a clue what was going on. In those early days, we learned the basics such as mission, organizational structure and function, history, the mechanics of contracting, program management, and security. About six months into this education, a notice came out concerning an innovation course. This was well before innovation became the overused term that it is today, so we immediately signed up. We looked forward to attending this course, but a few days later another notice arrived informing us that the course was cancelled due to lack of interest. In retrospect, that should have been a clue about what we were up against.

In some sense, this book is the course we would have liked to have taken at that time; i.e., something beyond facts and recipes that would help us to get the most out of the time we would give to government. It covers how we thought about problems in government, how we created an environment that drew the best people and ideas, and how we worked with people to get things done efficiently.

Bureaucratic Aversion to Change

If you are already in an environment that actively seeks new ideas and works very hard to implement them, then you are a very lucky person. Most government organizations are adverse to change and spend most of their time protecting money and positions. Even those that recognize the need to change in order to perform their mission often battle tenaciously over how and when that change should occur.

Machiavelli probably said it best: "And let it be noted that there is no more delicate matter to take in hand, nor more dangerous to conduct, nor more doubtful in its success, than to set up as the leader in the introduction of changes. For he who innovates will have for his enemies all those who are well off under the existing order of things, and only lukewarm supporters in those who might be better off under the new."[1] While Machiavelli was providing advice to rulers, resistance to change is independent of the scale of the issue and can occur from the lowest to the highest levels of government. That was certainly the case in the CIA.

Failing Forward Fast

As COTRs in research and development, these are some of the general issues that kept us up at night:

- We didn't know enough about a problem to derive a workable solution.
- Many possible solutions existed, and sorting them out required data that could only be acquired by contract.
- Contracting always seemed to take too much time.
- Schedule slips and failures were inevitable, except in hindsight.
- Priorities changed, making the problem we had been working on irrelevant. Key people left and their replacements had different views of how things should be done.
- There was too little funding, too much funding, or the funding was in limbo.

Uncertainty and risk are the norm in an R&D project that tries to solve an operational problem. Any of the above issues can fatally delay or derail your effort, and some of them are largely out of your control. We faced these and related problems in government for 25 years and concluded that besides judicious selection of problems and customers, your best protection against program failure is how you approach problems. The key is flexibility in

---

[1] Nicolo Machiavelli, *The Prince and Other Writings*, trans. N.H. Thompson, in *Harvard Classics Vol. 36* (New York: P.F. Collier & Son, 1910), 20-21.

how you work. Cast a very wide net at the beginning, use flexible experimental contracts to clarify issues, and revise plans frequently to reflect reality.

One of our colleagues called this process "failing forward fast," and we like the term for several reasons. First, it recognizes that failure is how success ultimately comes about. Revisionism and Monday-morning quarterbacking tend to obscure this fact, in part because the story always looks neater in the rear-view mirror. Second, when failure is managed properly and we learn its lessons, we move closer to the goal. We now know something that doesn't work and (if we did it right) why it doesn't work. Failure can provide insight that is not apparent otherwise, and allows us to more confidently make the next decision. And lastly, "fast" captures the sad truth that we only have a certain amount of time to get things done. A sense of urgency is important if you believe your time and energy matter. In government, this means keeping processes like contracting flexible and well-oiled.

You can think of failing forward fast as the technical version of kissing a lot of frogs in order to find the prince. It is messy, hard to predict, and requires a great deal of flexibility - all characteristics that are not usually associated with government contracting. But if you want to make the best of your time in government, you must try. Failing forward fast can be especially effective when you are trying to solve problems that are murky and have no obvious solution; in other words, the hard stuff.

A Quick Tour through This Book

We've divided most of what we have to say into 10 main chapters, beginning with a bit about the CIA and how to think about work environments and ending with some thoughts on retirement. Some of the ideas that we will suggest might seem over the edge to you, or inappropriate for your environment, and that is just fine with us. Use only what you find interesting and discard the rest.

Working at CIA

Most of our more enlightening experiences with bureaucracies came from our time at the Central Intelligence Agency. Since CIA is often misunderstood, we will give you some sense of the organization and introduce you to one of the people who influenced us.

Work Environments

In our careers we have worked in some really great and some really not-so-great environments. We will discuss the characteristics that we think are most important in an effective organization and introduce you to the concept of a virtual organization in government. While every bureaucracy has an organizational chart, you should not limit yourself to the boundaries that it implies. Find talent and allies wherever they might be and make them part of your team. By creating your own virtual organization that does not respect the org chart boundaries, you can do better than whatever management would otherwise provide.

Personal Skills

What you are able to accomplish in government has as much to do with you and what you do well as anything that the government brings to the table. You are not going to get to a high level of productivity simply by winging it and ignoring your flaws. In this chapter we will discuss a number of skills that we found to be particularly effective. Presentation skills, memory, quantitative abilities, learning, and planning are some of the talents that you will need to develop to a level beyond what your peers possess. Ignoring such skills means that you are willing to leave your fate in the hands of others, and that can be tragic. We have known many government employees who essentially gave up in mid-career and started to count the number of years they would have to endure before they could retire. You don't have to be one of them.

Self-Deception

You will make many decisions during your career such as what is important, who is competent, and which technologies are ready for prime time. The quality of those decisions in large part will depend on how well you can avoid deceiving yourself. We all suffer from biases, and effective program managers learn to recognize them and take active measures to avoid their effects. Those of you who are thinking: "I don't have a problem with this" could be some of the worst offenders. We will show you a few things about your brain that just might change your mind.

Program Management and Contracting

Unless you work full time in a laboratory, program management (PM) and contracting are at the core of what you will accomplish. You cannot create an effective environment in which to fail forward fast without strong leadership and management skills. Government PM and contracting courses tend to cover the legal and procedural processes that you need to know, but unfortunately, that is not nearly enough to be an effective program manager and COTR. One of the most dangerous things in the world just might be a newly-minted COTR with a couple of courses under his or her belt. In two chapters on these subjects, we will try to provide you with some of the more critical insights that we learned the hard way.

Some Things You Should Never Do

Government can be an unforgiving place, and it is important to know what can get you into real trouble. We're not referring to time card fraud or poisoning your boss, but rather behaviors we have observed that have negative long-term consequences. For example, a reputation for not being able to keep your mouth shut is going to impact the kinds of problems you are assigned. While the

15

number of such negative behaviors is probably very large, we've winnowed our list down to a dozen things you never want to do.

Working Your Way out of a Mess

You will probably find yourself truly stuck and frustrated on many occasions. These moments are important times for you, not only because they impact your progress, but because they help to define how contractors and co-workers view you. Your stock goes up when you elegantly resolve a problem that everyone believes is hard. In a failing-forward-fast environment, you do not want to always depend on the system to solve your problem. We'll show you some of the methods that we have used.

Staying Healthy

A failing-forward-fast environment can be very stressful because you've got a lot of balls in the air and people are depending on you to deliver. To help you cope, we have included a chapter on some ways that you can stay a bit healthier in the job. We did not always take our own advice when it came to healthy living. In our early days, we could work our way through an all-you-can-eat dessert bar and considered it a wonderful dinner. That kind of behavior will catch up with you eventually, however. In this chapter we will show you some ways to prevent your own meltdown.

Retirement

Some of you are probably planning to die at your desk, but we hope that most of you will have had enough at some point and want to retire to something else. For that reason, the last chapter is on retired life after government. We are not sure that we have anything unique to say about leaving it all behind, but given how often people screw up retirement, we thought we should provide our perspective. Retirement is no reason to forget failing forward fast and the planning and execution skills that it engenders. They can help make retirement as interesting and productive as any work career could possibly be.

Getting Started

We are big fans of puzzles in part because they can have a lot in common with failing forward fast. Try to solve this one.

Given the following maze, determine how many solutions there are and prove that they are the only solutions. Include only the solutions that do not require backtracking. You have 15 minutes.

END

START

17

We chose this specific type of problem because the process that one uses to solve it has much in common with failing forward fast; i.e., find what doesn't work and use that information to move forward. Essentially, the maze defines a number of possibilities that you study and reject or keep.

Look at the maze again and start eliminating the dead ends. These are the portions of the maze that are closed, so that once you enter them, you cannot get out except by backtracking. These dead ends cannot be part of the final solution. For example, here is the maze after we blacked out roughly half of the dead ends with a felt-tipped pen:

END

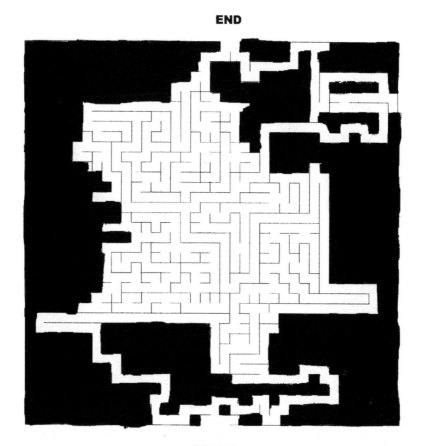

START

18

And here is the maze with all of the dead ends eliminated.

END

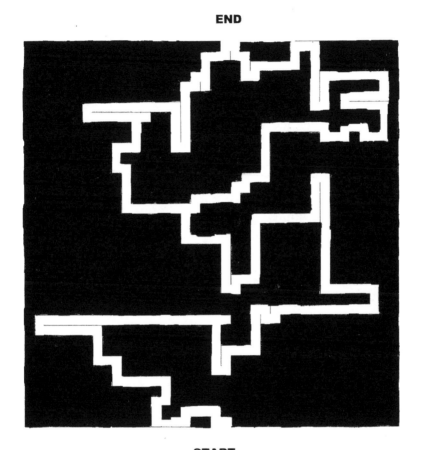

START

At this point it should be clear that there are only two paths that meet the criteria.

We showed mazes like this to dozens of groups of new Agency employees and their reactions were rather telling. Many started to solve the maze even before we told them what the problem was. Some of them thought the trick was to start at the end and work backward, but we pointed out that this was no different than turning the puzzle upside down, so how could that strategy possibly be better than starting at the beginning? When we

19

interrupted someone who was busily tracing a path, we would ask them if they believed that their method would answer the question. Once they thought about it a bit, they realized that it wouldn't. And finally, people never seemed to work together. It was as if they thought they were still taking tests in high school.

As you might suspect by now, these are some of the same types of mistakes that new COTRs in the government make with their programs. They jump in without knowing the real problem. They work on solutions that will ultimately not address the problem. And they are reluctant to ask for help from their colleagues. These and many other issues dilute the effectiveness of a COTR over his or her career. The following chapters will keep you from being a victim of such self-limited thinking.

CHAPTER ONE

Working at the Central Intelligence Agency

CIA..... Your formula for a career that makes a difference.

(Baltimore Sun want ad on June 2, 1985)

This book is about working more effectively in bureaucracies, and we are going to use the CIA as our example since it is the organization we know best. Unfortunately, what you the reader know about "The Agency" at this point is probably inaccurate. Let us tell you a bit about this often-misunderstood place where we grew up.

Welcome to the CIA

Being a CIA employee does not mean that you know everything there is to know about the CIA. Authors of books on the Agency typically have experienced some part of the organization, but it is rare to find someone who has seen it all. Part of this is due to the compartmentation or isolation that is needed for successful clandestine operations. But part is also due to the fact that the Agency is not homogeneous. The directorates that make up the organization have very different cultures and one cannot assume that experience gained in one translates well to another. We were fortunate to see a bit more of it than most because, as researchers, we had customers in all of the major components.

The CIA is a complex organization with a large and challenging mission. We were told by one senior legal official, for example, that at any given time, the Office of General Counsel (OGC) was dealing with on the order of 15,000 issues related to missions. Even if this number was exaggerated a bit for effect, it still suggests that no one knows it all. Not even the Director of CIA could possibly be familiar with the details of every issue that the CIA is grappling with. Keep this in mind the next time you hear some pundit talking on the evening news.

The CIA is the same as most bureaucracies in many ways. The organization is generally dominated by managers rather than leaders and has its share of suffocating rules and procedures. It is also plagued by common bureaucratic ailments such as a lousy tooth-to-tail ratio, an obsession with rice bowls, excessive political correctness, and an unhealthy fear of failure.

On the other hand, the CIA is different in some important ways from most other bureaucracies. One is its secure environment and the special limitations on sharing that security entails. Communications and trust are always problems in bureaucracies, but the issues are compounded when you add the complexity of classification. Another difference is that CIA employees are generally above average. Not just anyone can join the CIA because the extensive interviews, tests, and background checks weed out most individuals. This selection process is critical since employees have to understand and conform to a complicated set of regulations that both enable them to break some laws of other countries and stay within the bounds of the laws of the United States.

What the CIA does and how it does it have been distorted for decades by movie makers, novelists, and television writers. CIA exists for several reasons: to collect foreign intelligence by various means; to provide analysis for a variety of government customers; and to engage in covert action at the request of the President. Unlike the Federal Bureau of Investigation (FBI), the CIA is not a law enforcement agency.

Despite what you may have seen or read, the CIA does not do whatever it wants to. It is a service organization that provides help to a number of customers, including the President. Unfortunately, the CIA just can't be as transparent as some folks would like it to be, and that leads to interesting speculation.[2] Confidential sources and methods are the only reason the CIA can gather some of the key information that its customers request. Otherwise, disinformation and counterintelligence efforts will ensure that the needed information is never found or is deliberately corrupted in some fashion.

Because of its worldwide mission and the variability, sophistication, and resources of its targets, the CIA hires people with diverse backgrounds, interests, and experiences. This is true even within R&D because intelligence problems are technically diverse, and even solving one rarely involves a single discipline. A simple problem involving an implanted sensor, for example, might require a variety of expertise such as concealment, deception, health and safety, legal, and environmental issues, as well as the obvious technical problems of how a device does what it does, how small it must be, and how it should be powered. We were expected to use whatever resources that were available to find solutions to these multidisciplinary problems. As a result we were

---

[2] In 25 years we were unable to find any evidence that the CIA either keeps aliens in hidden facilities or has a portal to Mars. We mention this only because we have been asked on several occasions.

constantly pushed out of our comfort zones and worked with many employees and contractors who saw the world in different ways. While it is hard to believe, we were sometimes required to deal with chemists, even though their stuff never worked.[3]

The People Who Helped Us

We learned many of the ideas that we describe in this book from a number of very talented people. One of them was our first boss, Bert. When we met him he was a newly-minted branch chief, a position he got despite not being well liked by upper management. Bert was honest to a fault. He never sugarcoated anything, and this often put him on the outs with upper management. We really liked the fact that he said what he thought. A couple of stories will help you to understand him and how he influenced us.

Early on, we were asked to take over an on-going program involving radio-frequency imaging.[4] Things went well until there was a build phase of the program in which a subcontractor had a major portion of the work. The subcontractor quoted a price for their work that was double the initial estimate. This higher price was still cheaper than any other quote we received, so we needed to find more money in order to continue. The cost growth would be on the agenda for our next branch meeting.

At this meeting, another project manager briefed a different program before us. Several questions came up and Bert told the project manager to meet with him afterwards to formulate a solution. This made us a little apprehensive since we needed a significant portion of the branch's budget to fix the problem.

---

[3] Just kidding, Bill and John and Joe.
[4] In the private sector of 1986, a person with our level of experience would never have been put in charge of a multi-million dollar project such as this radar effort. CIA put people in positions of significant control early in their careers, and this helped to make them better program managers in the long term.

When it was our turn, we explained the problem and what it would cost. Bert then asked if we had a technical solution. We said that we had, and Bert replied: "Well, do it. You have more experience than I do in this area." This set us back a bit. Our new boss had confidence in us and backed it with money - something we hadn't experienced that much in previous industrial research jobs.

Bert always stood up for his people. At one point, we were asked to help solve a hostage rescue problem, and came up with a method for using radar as part of the solution. Bert took the idea to a senior staff meeting and sold them on funding the work. Several years passed before we found out what really happened at that meeting. One of the senior scientists said that he thought the idea was good and that he should do it. Bert became unglued and said that the right thing was for us to carry the idea to fruition and not have someone else get the credit. Bert knew that the one who came up with the solution was sometimes forgotten when the project was done and credit was given. In the end, we kept the program. Unfortunately, the encounter created several invisible enemies, and looking back, this animosity explained a lot at promotion time. No one ever said that life was fair in government, but Bert did his best to protect his people.

One of the first promises that Bert made was that he'd stand up for us if we were taking flack in a meeting with management, but later and in private, serious talk would be needed. Bert stuck to this policy for as long as we knew him. He had a long career in the Navy and the Agency and made it a point to treat people the way he wanted to be treated. This was one benefit of his becoming a branch chief later in life, in that he worked to do the right thing rather than trying to get promoted. Bert instilled a sense of pride in us and challenged us to be better. Several of us who worked for Bert were later promoted into the Senior Intelligence Service (SIS) and that is why we often referred to him as a kingmaker.[5] He never became a king himself.

---

[5] The Senior Intelligence Service is the CIA's equivalent of the Federal Government's Senior Executive Service.

Bert had only two fears in life. One was his wife, Susie. We remember Bert calling in sick one day and explaining that he had to do emergency wallpapering. It was a matter of life and death. We asked him what emergency wallpapering was, and he explained that his wife said she'd kill him if he didn't do it now.

Snow was Bert's other fear. He had lived in the Washington DC area for many years and knew how badly DC drivers handled snow. A couple of snowflakes were enough for him to run down the hall telling everyone to go home. We all needed to beat the maniacs and he knew that there was no point in becoming part of the problem. He always liked to remind us that the government would not pay to repair your car if you had an accident while commuting. You can't argue with that logic.

In the end, Bert and several other managers like him taught us to believe in ourselves. They allowed us to create our own management styles and grow within the organization. We were very lucky to meet them early in our careers.

CHAPTER TWO

Creating a More Effective Work Environment

You might be God's gift to engineering, but if you cannot figure out how to create an organization around you that helps to use that talent, then you will never fail forward fast.

Most employees would agree that government is not inherently a failing-forward-fast environment. "Failure" is usually considered to be a dirty word in the government; "forward" is often ambiguous; and "fast" is rarely part of the vocabulary. Many talented scientists and engineers join a bureaucracy with high hopes of doing great things, but they are ground up by the system. In this chapter, we will discuss some of the characteristics that we think are important in an effective environment.

What Kind of Organization Are You In?

There is a straightforward test to tell what kind of technical environment you work in. Simply answer this question: do people in your organization talk mostly about people, the system, or ideas? In an effective R&D environment, the emphasis should be on ideas. Excessive gossip and complaining about the bureaucracy and its people are toxic.

Most managers recognize that there are elements common to effective environments. These include a mission that everyone understands and supports, as well as honest and trustworthy management. Such basic attributes are important, and you can find more about them by reading almost any management book. But there are others that we think are just as important when you need to fail forward fast. They help to make sure that ideas flow, that they receive sufficient critical review, and that the right work gets done to implement them:

- How are decisions made?
- How quickly can you react?
- Do people willingly help each other and share ideas?
- How are contractors treated?
- Do you have people you can count on?
- Is there much laughter in your work environment?
- Who charts your career?
- What does your boss believe his or her job really is?

Answers to questions like these should provide some indication of the effectiveness of your R&D environment and suggest ways to improve it. We'll consider each of these in some detail.

How Are Decisions Made?

Every organization constantly makes decisions, and how those decisions are made is a key both to morale and to your ability to get things done. Ideally, decisions are made on the basis of rational thought and incisive questioning. Unfortunately, personalities, emotions, and biases also play a role. Some of the times we have seen bad decisions made were when:

- Sexual attraction was involved.
- The decision maker gave in to whining or threats.
- The decision maker took the easy way out of a problem.
- A "not invented here" attitude prevailed.
- Vested interests dominated the decision.

It can be discouraging to work in an office where whiners win and data is ignored over "gut feelings" and other weak reasoning. The best you can do in such circumstances is to try to make sure that decisions you make are based on solid data and reasoning rather than biases.

Be especially wary of offices that rely too much on committees to make decisions. In our experience, committees that deal with serious issues often focus more on perception or on protection of vested interests than acting on the best ideas. In research funding, for example, committees typically ensure that everyone gets a piece of the pie because that is the least controversial solution. Unfortunately, the least controversial solution is not always the best solution or even a good solution. Think very hard before you form a committee. In many cases, one fair-minded individual combined with a sanity check by a few others can do a better job than a group, if that one individual simply takes the time to talk with the right people. Few people will work as hard on a committee as they will as an individual.

Some might see this view of committees as cynical, and point out that there is considerable advantage in having many people interact simultaneously. This can be true and we admit to being on some pretty good committees during our careers. But most of the serious committees we have been on were not that good. Management often included far too many people on a committee, spent too little time getting the right people, and ignored the dampening effects that committees have on the willingness of some people to speak out. This last item is often dismissed, but some folks will only tell you what they are truly thinking when no one else is around. You have to create an environment of trust to let that happen, and those conditions often do not exist in a committee.

Do not confuse committees with more effective methods of working with others such as in well-designed teams. Committees are often selected to protect vested interests. In an effective team, on the other hand, those interests are discarded in pursuit of the best solution to a problem. Working with the right people usually leads to better solutions than working in isolation. As we noted before, solving a problem for the government is not the same as doing a math problem on a high school exam. So why limit yourself in the same way?

How Quickly Can You React?

Government works on the basis of a fiscal year that runs from October 1st through September 30th, and if you accept this schedule as your only schedule, you will not have the flexibility and opportunities that you need to fail forward fast. If a problem is discovered, and the only approach you have is to "put it in the budget for next year," then you are not going to get much done. You need ways to start early on projects, irrespective of the fiscal year process. We will discuss this in much greater detail in the later sections, but here is one example to show how we addressed the issue.

In the early 2000s, we decided that we needed to be more agile in order to make better programmatic decisions and help our customers with their problems. While some answers can be found with a bit of personal homework, others require specialized expertise and equipment. Our goal was to be able to pick up a phone and have an experiment run or a question answered that same day. The key was having on-demand access to a sufficiently diverse group of cleared scientists and engineers who could help us, either through analysis or experimentation. Some of our larger national laboratories were perfect for this capability, so we worked out an acceptable way to park funds in the Department of Energy to use on an as-needed basis.[6] This meant that we researchers probably had the fastest response time of any part of the organization for a wide range of problems - not what most people would expect.

While we built this capability primarily to provide us with data on which to make contractual decisions, the first use of this mechanism was in the early days following the September 11, 2001 attacks on New York City and Washington, DC. We needed to quickly bring together technical experts on various types of weapons to assess current intelligence and prepare for possible additional attacks. Having an existing agreement and funds in place to do work cut the time to almost nothing. Others in the Intelligence Community later created their own version of this idea.

---

[6] "Park" means to put funds somewhere in anticipation of use in the future. Bureaucracies hate the idea, but some form of fiscal flexibility is absolutely necessary in a failing-forward-fast environment. In our case, we were able to successfully argue our position by limiting the scope of such studies and placing some controls with DOE management. There are times when such tools are considered reasonable and necessary and others when management considers budgetary appearances more important than getting things done. In our case, we were fortunate to have been at the right place at the right time.

Do People Willingly Help Each Other and Share Ideas?

We recall a time when we received an unusual call from a colleague. She was just put in charge of a special unit that was responsible for R&D related to weapons proliferation issues, and she wanted to know what our secret was - why folks came to us for help. We told her we weren't sure, but suspected that part of the answer was that we would always try to help. And if we couldn't help, we were at least sympathetic listeners.

You might think that you are too busy to deal with other folks, but that is the wrong attitude. During WWII, for example, few people were busier than Vannevar Bush, the director of the Office of Scientific Research and Development (OSRD). OSRD was responsible for coordinating scientific work for the military, addressing a wide range of problems that included radar, the proximity fuse, more effective medical treatments, and better vehicles. In his book, *Pieces of the Action*, Bush describes an occasion when his secretary told him to see a particular visitor to the office. This visitor was an old man from the Blue Ridge Mountains who had driven to Washington to make sure that the government heard his idea for trapping submarines. OSRD received a lot of ideas on the subject, but Bush treated the man with great respect, listened to the idea, and admitted that they might be working on that very idea at the moment, but could not tell him about it because of secrecy issues. In our mind, that was a class act and a behavioral model to emulate.[7]

You need to be willing to help others even if it is a significant drain on your time. In addition to being the civil thing to do, helping others is a critical part of keeping a network of people

---

[7] Vannevar Bush, *Pieces of the Action* (New York: William Morrow and Company, 1970), 138.

together, and particularly creating a network of people who owe you favors.

As part of your efforts to help others, you should try to give away every good idea that you have, because hoarding is the surest way to slow you down and keep you from failing forward fast. Basically, any idea that you can convince someone else to pursue is one more idea that could produce results. Your hands are free to move on to other things.[8]

We used to say that the lowest form of the art was to do something yourself with your own allocated funding, and that a higher form was to convince someone else to do the work for you using your funds. But the highest form of all was to get someone else to do the work with their funds. This explains in part why we thought it was important to spend time helping technical personnel in other agencies who had their own funds and problems similar to ours.

How are Contractors Treated?

Contractors should be treated as equal members of your organization. They are not part of the team just to feed your ego, to be bossed about, or treated like peons.

While this might sound a bit harsh, some COTRs really do treat their contractors poorly. We recall one contractor who told us about his encounter with a particular Agency COTR who had just taken over a program. During a break at a design review, the contractor asked the COTR if he was learning anything new at this meeting. The COTR response was: "I don't come to these

---

[8] Clearly, you shouldn't just toss ideas about, but rather try to match people to ideas. A new person, for example, might benefit more from an idea that can be brought to closure in a short period of time rather than something that might take years to resolve. In this way, you can help them to more quickly gain a reputation for being able to finish things.

meetings to learn anything, I just hope we finish early so I can go play tennis." Even the term "complete idiot" doesn't adequately capture that COTR.

You are not going to get very far without good contractors. Even if you really are the Albert Einstein of your field, you will accomplish much more if you can effectively engage good contractors in your pursuits. Keep in mind that contractors are often the historical memory for a particular technology or operational concept. You need them on your side if only to educate you about what came before.

All of this assumes, of course, that you have found the really good contractors. We have known many that we were honored to work with and treated them accordingly. They rarely disappointed us. More importantly, they wanted to work with us. This mutual respect helped us to create and maintain effective teams, something that became increasingly important as more of our contractors moved toward an "engineers are replaceable cogs" mentality.

Do You Have People You Can Count On?

Life in the government can be pretty lonely if you don't have people you can count on. You need them to cover your back, to provide confidential advice and guidance, and to occasionally do things that you cannot accomplish yourself.

This protection is necessary in part because spiteful people exist even in the best of organizations. There will always be folks who do not like you and will actively try to sabotage you and your efforts. The list of these miscreants will likely grow the longer you stay and the more decisions you make or are blamed for. So, you need friends who will watch your back. They will see and hear things that you won't, and in some cases, they can help to diffuse the friction.

Do not be bashful about seeking out more-experienced people to advise you. The right advice from folks who have been where you are can save you time and energy that you could be using to solve technical problems. While organizations often have formal mentoring programs, you are probably much better off establishing relationships outside of the bureaucratic structure. The program overhead really doesn't contribute much, and any mentor who is in the program because mentoring is one of their performance goals is probably not the person you want. Also, you need more than one mentor in order to get different perspectives.

In our own careers, we found curmudgeons to be particularly reliable mentors. While the term "curmudgeon" has some negative connotations (for example, it often refers to someone who is easily annoyed), we viewed most of them fondly. For us, the key to curmudgeons is that they tell you what they believe to be the truth. Curmudgeons don't worry so much about being sensitive to your feelings. The good ones survive in part because effective management knows that it needs to hear the blunt truth at some point. Curmudgeons are a necessary evolutionary (and sometimes revolutionary) component of an organization.

If you want to have people on whom you can count, then learn how to keep a secret. It does not matter whether you work in a classified environment like the CIA or in an organization that deals only with unclassified matters. Everyone has secrets to keep and if you are known as someone who can keep your mouth shut, then trust starts to form. Many people cannot keep a secret and you need to be very careful what you say to them. Even some of our elected representatives are willing to spill secrets if it suits their purpose. If you are one of the many people who cannot keep their mouths shut, you will have to change in order to be most effective.

35

Is There Much Laughter in Your Work Environment?

A good sense of humor was a critical skill to have when we first started at the Agency. Folks were constantly joking with one another, and it was generally wonderful. One of our better ideas back then was for a fictitious company that we called Katzenjammer Industries, or KJI. KJI was a take-off on a comic strip that ran during the first part of the 20th century.[9] The idea started with a secretary who once referred to us as the "Katzenjammer Kids," who were the chief source of mischief in the cartoon. This fictitious corporation gradually became part of our division culture, and terms like "a Katzenjammer solution" became commonplace.

KJI's main activity was making silly gifts for members of the office. The gifts were typically take-offs from stories or comments that an individual made at some point in their career, so it always paid to listen carefully to what others said. For example, our co-worker and mentor, Thad, was an enthusiastic DIYer. He made the mistake of telling us about the time he accidently glued his foot to a linoleum floor. Needless to say, we brought this up many times in various situations during his career. When Thad retired, KJI gave him a plaque that was a rubber foot glued to a linoleum floor tile.

We eventually had to shut KJI down because the word got out and senior people would come to us asking to make gifts for people we didn't know. Needless to say, they didn't understand what made the idea tick. The gifts only worked because they showed an understanding and appreciation of the individual.

In addition to helping morale, humor causes you to break thinking patterns and can help to stimulate new ideas. If you're stuck on a

---

[9] Wikipedia, *The Katzenjammer Kids*, https://en.wikipedia.org/wiki/The_Ka tzenjammer_Kids (accessed July 8, 2015).

problem, you need to laugh and come back to the problem from another perspective. We've often used humor during brainstorming sessions when we were stuck and no new ideas were forthcoming.[10] Laugh, and then try again.

We were fortunate in being able to meet R. V. Jones during his visit to the Agency in the early 1990s. Jones was a physicist during World War II for whom the Agency (under Jim Woolsey) created an award for "scientific acumen, applied with art, in the cause of freedom." Jones was a very strong proponent of the art of practical joking.[11] We asked him about the importance of humor in his efforts in World War II and he replied that humor was absolutely essential in keeping people from going crazy. He and others were working 24 hours a day on solutions to life-and-death problems. Humor and practical jokes were essential to breaking the tension and making it easier to find answers to difficult problems.

Even in today's world, appropriate humor does not have to hide. We think that we played some pretty good tricks on our friends and that they appreciated the effort. One of them involved our friend Roger. At the time, we were serving a tour as a staffer in the Office of the Deputy Director of Science and Technology (O/DDS&T). It was not our best time. We were better suited to technical work than budget submissions and congressional crisis response.[12] But we learned a lot. Roger, on the other hand, had been recently promoted and was happy in ORD in his role as a branch chief - the same level that we left to join the staff. He was scared to death that his career board would soon stick him with a staff job, and made it clear to all that he had no interest in such a "growth assignment." So we thought we would have a little fun at

---

[10] We use the term "brainstorming" only in the sense of coming up with useable ideas. We do not believe in the formal rules that limit criticism.

[11] R.V. Jones, "The Theory of Practical Joking - Its Relevance to Physics," *The Institute of Physics Bulletin* 8 (June 1957): 193-201.

[12] Staff work is not for everyone. One of our colleagues at the National Reconnaissance Office, for example, often referred to his time as a staffer as his "bottom-feeding period."

his expense. Figure 1 is a copy of the memorandum that we forged, welcoming him as our replacement. The memo was "signed" by the DDS&T. We think Roger figured it out about mid-way through the second paragraph, right after he retrieved his heart from his throat.

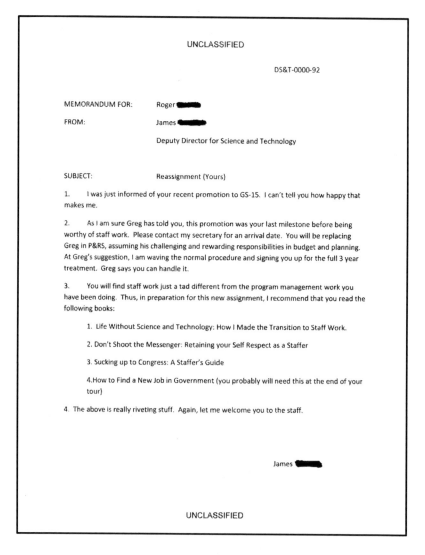

Figure 1. A Bogus Memo from the Boss

Kidding with colleagues is one thing, but for real excitement, you need to try kidding bosses. At one point, for example, we were required to submit copies of briefing charts and a biography for anyone giving a one-on-one presentation to the Deputy Director for Science and Technology (DDS&T). Since these documents were normally edited to death prior to reaching the DDS&T's desk, we decided to add a joke that would surely get pulled out by our boss or someone else in the review chain before the DDS&T would ever see it. The person who would be doing the briefing was one of our employees, Matt, who worked in telecommunications. We added a section to Matt's biography that referenced his invention of the number "1", and noted that this greatly increased the efficiency of the Internet since they could only send the number "0" prior to Matt's invention. Matt never knew that we included this in his biography. We forgot about the note until Matt came to us after his presentation and said that when he entered the front office conference room, the first words spoken were: "so you're the guy who invented the number 1." The briefing went well.

We also occasionally did classified cartoons as a way of connecting with some of our coworkers (Figure 2).

Figure 2. An Early Classified (and Redacted) Cartoon

To our knowledge, they are the only classified cartoons in existence. They will be probably be declassified and submitted to the National Archives at some point in the future, so keep an eye out for them.[13]

During our time at the Agency, we saw many of the offices become increasingly humorless. We hope that a better balance can be struck in the future, since we believe that humor is absolutely critical for a failing-forward-fast environment. The most effective teams that we knew laughed and joked among each other constantly. You can gauge an office's attitude to some extent simply by listening for laughter and observing items like cartoons on cubicle walls.

Who Charts Your Career?

You want to chart your own career rather than have someone else or a committee decide for you. We recall when we wanted to spend a few years at a national lab to recalibrate and delve deeply into robotics. Our current office director did his best to discourage this with a self-serving argument that our careers were better served by staying in place working for him. We chose to ignore his advice. We left for three years, wound up working for a manager at the lab who eventually became our boss at CIA, and got promoted into the Senior Intelligence Service while still on this assignment. We clearly made the right decision.

Occasionally you will be given bad advice with the best of intentions. No one is perfect, least of all your boss and the committee that advises you on your career. But beyond that, people can give you bad advice about your career for a number of

---

[13] If you think this was a frivolous use of taxpayer funds, we can assure you that there is much in those archives that is less useful and less productive than those cartoons. And the people who drafted that junk were serious, rather than - like us - just trying to boost morale. You've got to loosen up to be at your best.

reasons. Sometimes, they really don't care and just want to get you out of their office. In other cases, they are just selfish and are going to give the advice that most benefits them (as in the previous example). And finally, bureaucrats have in general become increasingly reluctant to say what they really believe.

The degree to which the government has swung toward a softer "everyone gets a trophy" approach is harming its ability to get things done, and this is particularly true about career advice. When we started in the Agency, our bosses were for the most part blunt, to the point, and honest. We appreciated that frankness since it allowed us to grow both as program managers and individuals. Perhaps things will get better, but in the meantime, you should take that performance review with a grain of salt. Most people have a pretty good idea of what characteristics and skills they need to work on, and if you have followed the other lessons so far, you will have some trusted allies who will give you the feedback that the system will not.

What Does Your Boss Believe His or Her Job Really Is?

Bosses can have different views of what their jobs really are. Some believe that their prime directive is to get promoted. Others believe it is to muck in your business and make all the tough decisions. Both are far off the mark.

We like the answer a Directorate of Operations manager once gave us to this question. He said that the only point of a good manager is to ensure that good people continue to show up and move through the ranks. He understood that people are hired to do the mission work and that managers need to know when to get out of the way and let them do it.

You should ask your boss this question some time when you get him or her out of the office. If your boss won't answer, then you will have to watch for actions that broadcast his or her management motivations.

Virtual Organizations

Chances are that you will not find all of the skills and people that you need within your own division or office. You have to look outside and engage components that cross organizational lines in order to get some of the help you need. One way to think about this process of reaching out is as a virtual organization that provides a path of least resistance through the bureaucracy. You are the leader, but not necessarily the boss in your virtual organization.

These organizational elements should include all of the resources you need to generate new ideas, test them out, adjust, narrow in on a solution, and implement it. They are the resources you count on to work through the bureaucratic processes associated with getting from ideas to field implementation.   Your virtual organization should include:

- Customers.
- Advisors (internal and consultants).
- Support (contracts, legal, accounting).
- Contractors.
- Access Agents (try to get someone you know and trust in every relevant office and agency).

Other COTRs should also be part of your organization.

If Figure 3 is how your organization looks to others, Figure 4 is how it should look to you.

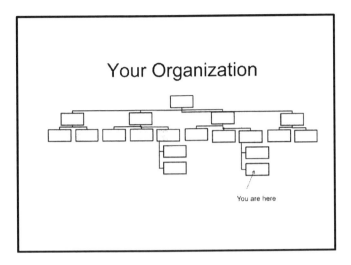

Figure 3. Your Organization

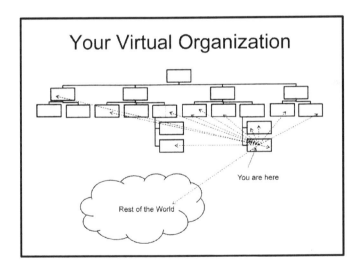

Figure 4. Your Virtual Organization

You will need years to form an effective virtual organization, but it all starts with the first people you meet. Use a failing-forward-fast philosophy to meet lots of new people, evaluate them, and find the ones who belong on the team. The Agency's tendency to move folks around within the organization and among other government agencies helped us to form our networks. For example, if a contracting officer we worked with for years in one office moved on, we would not write that individual off. In his or her new position, that individual often led us to even more useful contacts. As a general rule, try not to alienate any of the people you work with. Use the interactions to build trust.

One of our mentors once remarked that the only way to accomplish things in a bureaucracy is through splinter groups. The two of us in many ways were initially a splinter group of two and we ultimately expanded that into a larger and far more effective virtual organization.

All of this might sound a bit Pollyannaish, particularly if you work in a lousy environment. We understand that you are not going to be able to fix a broken bureaucracy on your own, but you can make your piece of it (your virtual organization) work better. Even in the worst office environments, there are always informal groups that form and survive. They might hate the job, and they might hate the boss, but they love the interaction/environment that they have created among themselves. Such folks want to work together and sometimes they make a good team that gets things done in spite of the general environment.

CHAPTER THREE

Personal Skills to Put You Ahead of the Crowd

Failing forward fast is in large part about you and what you bring to the table. Here are some skills you should develop.

If you want to be more effective during your government service, you have to get beyond the idea that "winging it" is a strategy. You will be more successful if you constantly work to improve your skills and prepare well. Not only will you make better decisions and attract better people, but you will get noticed and find that better and more important problems and opportunities will come your way. In other words, you will make your own luck.

In this chapter, we are going to describe several skills that we believe helped us do a better job. Some will seem obvious - your parents and friends probably told you some of the same things - but that doesn't mean that they are widely practiced in government.

These skills are well suited to a failing-forward-fast environment, and will give you an edge in your dealings with others.

Know Where You Are Strong and Where You Are Lacking

Do you know where your talents and weaknesses really lie? For example, are you good or bad at generating ideas, planning, problem solving, salesmanship, forming a useful team, being a team player, mentoring, program management, trouble shooting, tact, or diplomacy? Find your talents, refine them, and work hard to shore up weaknesses. If you find there is something you need to improve on, then find people with that talent and see what you can learn from them. Be honest with yourself. Remember, most people consider themselves to be above-average drivers.

Plan Early and Plan Often

The Boy Scouts got it right: "Be Prepared." You've got to have a plan to fail forward fast.

In fast moving environments, plans are obsolete shortly after they are written, but that is okay. Planning is less about producing the plan than getting the benefit of the deep thinking that such an activity requires. Plans should always be flexible and adapted as new information is gathered. No plan is perfect, but even a rough plan is better than doing nothing.

When you start a plan, there is always a flood of thoughts in your head. In many ways, it is too much to think about clearly and perhaps that is why many people do not plan very well. The trick that works best for us is to start at the end with the final goal rather than to start at the beginning.[14] While plans naturally flow from the present to the future, this alternative forces you to be very clear from the start about what you are trying to accomplish. Doing a plan backwards is a simple application of R.V. Jones's "The Other Way Around" principle in which one inverts a design in order to obtain an alternative solution.[15] We will discuss this approach later in Chapter 8 as a general method for solving problems.

Plans help to keep you focused, but they can also help you to anticipate and avoid disaster. For example, suppose that you ask for $3M to do some work, and are offered $1M. Do you blindly accept the funds and hope for the best when the money runs out? Or do you already know what you will likely have when those funds are spent and can temper the funder's expectations? If you have evidence that a smaller amount of money will never get you to a productive point, then perhaps the right answer is to say no.

You might be surprised at the impact a well-thought-out refusal can have. One of the more inspiring stories we've heard involved the program manager for a very large overhead collection system (aka a "spy" satellite). The device was the first of its kind, but Congress tried to cut the amount of funding for the system, making it far below the organization's estimates. This particular program manager was willing to walk away from the whole thing if Congress did what it was suggesting. In the end, Congress did not cut the budget. We don't know if it was courage or good poker, or both on the part of the program manager, but we clearly need more of it in government. You are at a real disadvantage in a negotiation if you aren't willing to walk away from the table. Remember that if Congress had succeeded in cutting the budget on the satellite program and the program failed to meet its goals as a result, the blame would still have fallen on the organization doing the

---

[14] Think of planning like describing a tree that you see in a forest. It is often easier to start with the trunk than with the individual branches.

[15] R. V. Jones, *Instruments and Experiences: Papers on Measurement and Instrument Design* (Chichester: Wiley, 1988), 460-474.

contracting. It is much easier to assign blame than to accept it in a bureaucracy.

Do not make the mistake of creating plans that cover only the technical portion of a project. A good plan should include the key issues that go beyond the technology. Take the customer for example. How should they be involved? When are the critical times that you will need them? What is plan B if a key customer is no longer available? You should have similar backup plans for any legal and funding issues that might come up. It also doesn't hurt to keep another COTR in the loop in case something happens to you.

Plans are not completed roadmaps to be followed mindlessly. A plan is a direction that needs to be updated as new information is obtained. If you don't change the plan to reflect new information, then you are willing to waste time and money on something that will not get the job done.

If you're building the same product for the tenth time, then changes to your plan will likely be minimal. But if you're building anything new and complex, you will likely have many changes to worry about. As we will discuss in greater detail in Chapter 5, the initial requirements that you receive for your project are likely incomplete and wrong to some extent. As you find out what's wrong, you will have to make adjustments. It is a bit like solving a crossword puzzle where you have some of the answers right, some of them wrong, and some of them remain unknown. In a failing-forward-fast environment, you must assume that you do not have correct answers to everything at the start, and that plans must change along the way. You can't make progress by continuing to do the wrong thing, and good planning habits can help you to avoid digging a deeper hole.

Develop a Good Memory

A good memory obviously helps a COTR since the easiest way to transport classified information is in your head. But a good memory can help in other ways, including making a good impression. How often, for example, have you been pleased when someone you met only once recalls your name long after the encounter? Some people have a knack for this, and there are many tricks that one can use to get better at remembering names and faces. Our only caveat is do not try to fake your way out of forgetful moments. Be honest. "I'm sorry, I remember meeting you at so and so's, but I am not recalling your name right now" is okay. "How are yooouuu?" (as you try to sneak a peek at their badge) doesn't work.[16]

A good memory can also help you make more effective presentations by eliminating the need to constantly refer to crib sheets. The ancient Romans gave public speeches that lasted for hours without benefit of notes. They often did this using the method of LOCI or Roman Room, which essentially associates the familiar (ordinary things in your life such as a room or a pathway) with the unfamiliar (what you are trying to remember and the order in which it occurs). For example, you know the layout of your own house or apartment. Relate the items you need to remember to familiar items in your home. Picture your front door as the beginning of your discussion and tie your starting item to it. Envision them together. Now move to the front hallway and relate your next item to it. Do this for all the points you need to remember. While this technique does require a little practice to be effective, you will be amazed at your ability to retain the information. We can successfully recall some presentations that we structured with the Roman Room method over 20 years ago.

---

[16] Those of you who are bad at recalling names can console yourself with the fact that your brain is largely based on human needs of 100,000 years ago when such information was less critical.

Many books are commercially available that can help you to retain information. *The Memory Book* by Harry Lorayne and Jerry Lucas is a good place to start.[17]

Learn to Be a Good Friend

Having a friend or confidant at work is important for many reasons, not the least of which is that you need someone to ground you, to tell you when you're wrong, and to explain why. Each of us, for example, went through several rough times when we were planning to quit.[18] During those times, one of us would challenge that decision and help the other to ride it out, usually by listening carefully, by arguing, and by getting the other person to laugh. It's amazing how different things can look two weeks later.

A real friend listens well, but does not necessarily agree with you about everything. A friend tells you when an idea doesn't make sense before you tell your boss about it. A friend tells you when you are being self-destructive, such as by avoiding a doctor or driving like a crazy person. A friend will also tell you those embarrassing things that others avoid, like those pants really are too tight or that homemade deodorant you are so proud of doesn't work.

You will not fail forward fast if you only have friends who agree with you. Many people stagnate because they don't want to hear anything negative. They tend to see all negative comments as personnel attacks. You, on the other hand, need to be thick skinned and appreciate criticism. As one of our grandmothers used

---

[17] Harry Lorayne and Jerry Lucas, *The Memory Book: The Classic Guide to Improving your Memory at Work, at School and at Play* (New York: Ballantine, 1974).

[18] Fortunately, we weren't looking at the same time or this book would have a different title.

to say, "It's one thing to be stupid; it's another to want to stay that way."

Good friends are hard to come by and unfortunately we don't have much advice on how to find them. You can't force real friendship. The best we can offer is that your chances of finding a good friend depend on getting to know people. You have to make the effort. In some sense, it can be a failing-forward-fast process.

Develop Good Presentation Skills

We've seen a lot of bad presentations in our careers. In some cases, nervousness was the problem (we all face that to some extent), but more often, the cause was poor preparation. If you don't know your purpose, your audience, and your subject, and don't spend some time putting a coherent message together, then you will be wasting time and making people uncomfortable.

Even if you have given a particular talk a dozen times, you always need to prepare for the next time. While the talk might have been well received by past audiences, the next audience might be different. What is their background and what are they expecting to hear? There is no one-size-fits-all talk.

Fielding questions is a critical part of a successful talk. Before a presentation, for example, we spent much of our preparation time anticipating questions that might arise. This forced us to think hard about the audience. Even with preparation, however, you will be surprised by some of the questions, and might not be able to answer all of them. When you are stuck, try turning to humor. Ask if you can buy a vowel or use one of your lifelines. If the answer hasn't come to you by then or you are still uncertain, tell the audience that you need to check on something first and will get back to them. Whatever you do, don't babble on when you don't have the answer. If you don't know, you don't know.

51

Giving talks is a good opportunity to recalibrate. One of our favorite games was to see how much we really knew about something that was in the talk. We would begin with a simple question that someone from the audience might ask, and then start to ask increasingly specific questions. Eventually, we would reach a point where we could no longer answer with confidence. These "deep dives" can be quite sobering, because no one really knows as much as they assume they do.

A danger of preparing too well is that you will get to the point where you no longer feel the words and you will start to sound a bit like an automaton. Don't let that happen. How you say something can be just as important as what you say, and giving a presentation is in some ways a form of acting. If you play your role correctly, it will help to hold an audience's attention, improve their comprehension of the message, and convey impressions that go beyond the words, such as your level of enthusiasm and your level of confidence.

Sometimes, you will not have the luxury of sufficient time to prepare. Speaking extemporaneously can be a challenge, but when you are told that the boss wants to see you in 5 minutes about so-and-so, or you are asked at the last minute to say a few words to inspire the group, don't panic. The best solution we have found for these last-minute presentations is to use whatever time you have to quickly create a mind map. Much has been written on the technique, but the basic idea is to diagram the main points and their relationships. The mind-mapping process serves several purposes: (1) it lessens the chance that you will forget something important, (2) it improves your chances of sounding logical and not babbling, and (3) it will calm you down. Figure 5 is a part of one we used several years ago.

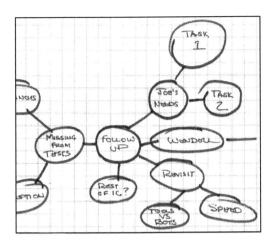

Figure 5. Portion of an Old Mind Map

As this example shows, mind maps don't need to be complex or pretty to effectively remind you of key points and relationships.

Learn to Think More Quantitatively

One of our all-time favorite scientists was Richard Feynman, a Cal Tech physicist who received the Nobel Prize for his work in electrodynamics. You might recall him from his many books including, *Surely You're Joking, Mr. Feynman!*.[19] He was also a highly-visible member of the review committee on NASA's Challenger explosion. Feynman was the one who put an "O" ring into a glass of ice water to illustrate the effect that temperature had on elasticity, elegantly demonstrating the risks that the launch managers had rationalized away.

Feynman was fortunate to have a father who - while not a scientist - had an unusual interest in science that he conveyed to his son. In

[19] Richard P. Feynman, *Surely You're Joking, Mr. Feynman!* (New York: W.W. Norton, 1985).

one of his books, Feynman described an example from his childhood in which his father was reading to him about dinosaurs. While most parents would simply read what was written, his father provided perspective and meaning. For example, if the book described the size of a *Tyrannosaurus rex* in feet, Feynman's father would try to put it in terms his son could understand. A 25-foot high dinosaur with a six-foot wide head as described in the *Encyclopedia Britannica,* for example, could stand in their front yard and look into the bedroom window, but its head would not fit through the window.[20]

Feynman's early exposure to such lessons probably helped to instill in him an innate sense of quantity and magnitude. This sense emerges strongly in his classic Cal Tech lecture, "There's Plenty of Room at the Bottom" which is credited by many as the start of the nanotechnology revolution.[21] In it, he describes the possibility of manipulating individual atoms. One of his conclusions was that it would be possible to write the *Encyclopedia Britannica* on the head of a pin using 1000 atoms as the equivalent of a half-tone dot. Thinking about the size of an atom and the size of the head of a *T Rex* head is not that much different. Decades later, IBM researchers isolated 35 Xenon atoms on a surface with a scanning tunneling microscope, spelling out the letters: "IBM."[22]

We believe that a strong sense of quantity and magnitude is an important tool in a failing-forward-fast environment. Your questions, your analyses, and your discussions will all benefit from a better feel for numbers. When we were involved in bomb detection, for example, we carried a 3x5 card that had held the vapor pressures and other physical characteristics of explosives typically employed by terrorists. These numbers were the key to

---

[20] Richard P. Feynman, *What Do You Care What Other People Think?* (New York: W.W. Norton, 1985), 12-16.

[21] Richard P. Feynman, "There is Plenty of Room at the Bottom," http://calteche s.library.caltech.edu/1976/1/1960Bottom.pdf (accessed 20 September 2015).

[22] Malcolm W. Browne, "2 Researchers Spell 'I.B.M.' Atom by Atom," *The New York Times*, April 5, 1990.

more effective ways to search for improvised explosive devices. The numbers guided our thinking and research, and were used to debunk a number of approaches involving vapor detection that were being funded by others. Because we paid attention to the reality of the numbers, we didn't waste time and resources on approaches that had no practical chance of working at the time.

Numbers can often become more meaningful when placed in a different context. One favorite historical example of this idea involves the migration of Americans to the west on the Oregon Trail from the 1840s to the 1870s. Most of the casualties along the trail were from diseases such as cholera and accidents. But how dangerous was it? One way to think about the danger is to cite the statistic that 20,000 to 30,000 travelers died during that period. That is large number for a population that was only a few tens of millions at the time, but perhaps a bit hard to relate to today when our population is now over 300 million A better way to think about it might be to note that approximately 200,000-300,000 pioneers made the trip, meaning that one in ten travelers died. And perhaps an even more insightful approach would be to recognize that those deaths represent a grave roughly every 100-200 yards along the 2,000 mile trail.[23]

Such tricks can help you greatly when you have to present technical topics to audiences that have few technical skills, such as in an agency like the CIA where many managers are technical lightweights. At one point during our careers, for example, the Agency was headed by a DCI who did not understand the difference between analog and digital communications. Since communications were at the heart of some Agency activities, this was a problem. Several senior scientists were tasked with correcting this problem, hopefully by using quantities and descriptions that the DCI could relate to. What an opportunity!

---

[23] Oregon Trail Interpretive Center, *Trail History, FAQS, and History Bits*, http: //www.blm.gov/or/oregontrail/ (accessed April 18, 2015).

Learn to Do Back-Of-The-Envelope Calculations

On one of our visits to a remote national lab, we marveled at the good fortune of the lab workers who never faced any significant traffic jams or other commuting problems. One of the lab managers who hosted us then commented that a minute of commuting was equivalent to an 8-hour day. Our immediate reaction was "that can't be right." But then we started thinking: get in a car, drive for one minute, get out, work the day, get back in the car, drive for one minute, and get out of the car. Figure on 50 weeks per year, times 5 work days per week, times 2 minutes per day, and you arrive at a bit more than 8 hours.

This example is one of the simplest cases of a back-of-the-envelope calculation. Physicists sometimes refer to this type of problem as a "Fermi problem," in honor of Enrico Fermi, who raised this problem-solving technique to an art form. An ability to do these types of calculations is one of the most valuable tools in your toolkit because of the insight it provides and the discipline it enforces in your thinking. While Fermi problems do not provide an exact solution, and they can sometimes be misleading (because of what is missing in the assumptions), they are a far better approach than simply taking on faith that something will work out. A simple calculation can often pull you safely back from fantasy land.

Fermi was a great teacher, and he took this way of thinking into the classroom. The classic problem that he and other physics teachers used to illustrate these calculations in their time was to determine how many piano tuners lived in Chicago in the 1950s.

At first glance, figuring out how many piano tuners live in Chicago seems like an impossible problem. Without cheating and looking up the number in a 1950s Chicago phonebook, it would seem that the best you could do is to pull a number out of thin air. But many of the problems you face in your jobs are like this, so it seems worthwhile to learn how to make such problems tractable.

While the piano-tuner problem was used by Fermi and others in the 1950s, its timeless aspect is the way that such a problem can be broken down into a series of questions that you can answer. For example, if you knew the approximate population of Chicago in the 1950s, roughly how many people were in a family, roughly how many families owned pianos, how often pianos were tuned, how many days per year a piano tuner worked, and how many pianos a tuner could adjust in a day, then you could get a ballpark estimate. Each of these individual problems is tractable to some degree, either by analogy, reference, or estimation.

The population of Chicago in the 1950s was around 3 million. Assuming 4 people per family would yield 750,000 families. It might be reasonable to assume that 1-in-10 families owned pianos at that time, creating a need to tune 75,000 pianos. If a piano is tuned on average once every two years, then 37,500 pianos would be tuned each year. If we further assume that piano tuners work 250 days per year and can tune 2 pianos per day, then Chicago needed 75 piano tuners in the 1950s.

The Chicago yellow pages a number of years ago listed 28 entities under piano tuning and repair. The number "75" is not "28," of course, but the closeness is somewhat amazing considering where we started. One tries to get into the ballpark with a Fermi problem rather than seek an exact figure. Sometimes, just knowing the order of magnitude of a quantity can help you to make better decisions. Fermi problems are also exceptionally helpful in giving you some sense of where the errors lie and how close your answer might be. If it turns out that you get a better estimate for the population of Chicago, for example, then you can adjust for it. Fermi problems are often less about the final number, and more about where the critical knowledge for a better estimate might lie.

The key to using back-of-the-envelope calculations is recognizing that the process is nothing like a guess. It is a way of calibrating your thinking by reducing uncertainty and getting a feel for a problem. In most good problems, you do not have all of the data and you don't have forever to accumulate it, yet you would really like to know how close you are and where the uncertainties lie. This description fits many of the problems that we (and likely you, too) have faced.

We used Fermi's methods to help design approaches to many problems that we dealt with. One example involved microrobotics, an area in which the Agency has some history. For example, one of the most famous microrobots is the Insectothopter, a dragonfly mimic that was built by the CIA in the 1970s. A picture of the device is shown in Figure 6. The device weighed about a gram and was powered by a gas generator that caused the wings to flap and provided some rear exhaust for propulsion. It was far ahead of its time, but (like most small flapping insects) performed poorly in high winds. Because of this deficiency and other issues, it was never used operationally.

Figure 6. CIA Insectothopter – One of the First Microrobotic Devices[24]

---

[24] CIA Museum - Central Intelligence Agency, *Experience the Collection*, https://www.cia.gov/about-cia/cia-museum (accessed April 17, 2015).

The interest in small robotic devices never completely died in the CIA, however, and during our careers we had to face numerous repeated questions about whether flying, crawling, or swimming versions of these small robots would ever become practical. Microcrawlers were of particular interest. The idea behind these devices was that they would be very small (roughly a cubic inch or less in volume), would move by walking or rolling, and could potentially get into interesting places without being detected. The conventional wisdom was always that these systems were not feasible, and the primary objection was that the power sources would not support the technology. This argument seemed reasonable to most folks. Power sources were almost always a problem, and an unwritten rule for many of us was to never start an operationally-focused program that depended on a new power source.

We were bothered by this reasoning about the limitations of the power source because it was never accompanied by even the simplest quantitative analysis. While there are many problems with microcrawlers that might make them a bad bet, we felt it was important to make sure we were rejecting them on a sound basis. Were power sources really the culprit? Perhaps Fermi could help here.

The general problem with tiny robots is scaling – how properties change as things get smaller. Let's say we take something like a cube and reduce its length, width, and height by a factor of 10. In this case, any property that depends on volume (such as mass) is going to decrease by a factor of 1000 and any property that depends on area (such as strength) is going to decrease by a factor of 100.

Since mass reduces by a factor of 1000 and strength only by 100, smaller robots can be relatively stronger than larger robots. That is why an ant can lift so much relative to its body weight. But there is also bad news that involves the power source. The amount of energy that the robot can store as size shrinks is drastically reduced because energy depends primarily upon volume rather than area or length.

A simple back-of-the-envelope calculation began to refine our thinking. How high could a battery be raised, for example, if all of its energy was used to lift it? If that distance was high enough, then perhaps the power source was not the limiting factor.

This was an easy calculation since energy is force (the weight of the battery) times the height moved. We selected an LR 44 battery for the power source since it had been used by others in microcrawlers.[25] A common brand of this 1.5 volt battery weighed about 1.9 grams, or 0.0186 newtons. Its charge capacity was about 105 milliamp hours which corresponded roughly to an energy output of 567 joules.[26] Equating this energy to the weight of the battery times the height, we calculated that the maximum height the battery could achieve was about 30,500 meters, or 19 miles! This result suggested that the limiting factor might not be the power source.

Once you have a handle on the simplest analysis, you can dig down further. For example, the battery is only a portion of the microcrawler and most of the weight that has to be lifted is not battery. In some of the prototype microcrawlers of the time, the battery was about 20% of the total weight. As a result, we had to reduce the heights we calculated by a factor of 5, but that was still 6100 meters, a pretty good travel distance for something that is on the order of a centimeter in length.

One should also factor in motor and gearbox efficiency (or its equivalent, depending on the actuator technology used). One of the more popular motor/gearboxes for small robots at the time had an efficiency of only 5%. This meant that the drive system only

---

[25] LR44, is an alkaline button cell that was used in a number of small electronic devices at the time.
[26] We roughly estimated the battery energy by multiplying the charge capacity times the nominal voltage.

converted $1/20^{th}$ of the energy it drew into useful work (going up, in this case). So we divided the height by 20, and still wound up with over 300 meters.

Depending on what you are trying to do, 300 meters could be interesting or not. But the insight that we gained concerning the problem of power was far better than the old answer that: "the power source is the problem." The analysis also suggested other options such as lighter-weight mechanical components and higher-efficiency actuators for achieving longer distances.

We think that this type of analysis is critical to failing forward fast and avoiding the problem of spending valuable time and resources on dead ends. But never assume that the answer you get from a back-of-the-envelope calculation is the last word about the subject. Remember that the payoff in these types of calculations is not so much the number, but rather the assumptions and thinking that went into it. If you get those wrong, then it is garbage in and garbage out. Do not lose sight of the fact that this process is primarily about gaining insight into a problem.

Despite the pitfalls that might occur, the bottom line is that you will probably make better decisions in a failing-forward-fast environment if you can quantify the options. You will have a better idea of what to explore and what to ignore. Unfortunately, you don't get good at doing Fermi problems by reading a book. You get good by finding any opportunity you can to practice the craft. In the end, it is a state of mind made better by continual use.

Read More

A good friend of ours once told us that if a reader got one new idea or insight from a book, it was worth the effort. Whether you are merely entertained by a book or helped by it depends entirely on whether you can spot the opportunities. For example, we recall reading Matt Ridley's *The Red Queen*, a book about human nature and its evolution.[27] The title comes from a statement made by the Red Queen in Lewis Carroll's book, *Through the Looking Glass*.[28] While describing her world to Alice, the Red Queen explains: "Now, here, you see, it takes all the running you can do, to keep in the same place." Ridley used the statement because of its similarity to Nature's need to adapt in order to survive. At the time we read this book, the Intelligence Community was struggling to adapt to a more modern world in which technology evolved very rapidly. Lewis Carroll's quote fit the circumstance perfectly because we were having trouble getting ahead of the technologies and had to run hard just stay where we were in terms of relative technical advantage. We started to use a Red Queen presentation slide that contained the above quote and a pen-and-ink drawing of the Queen. That vugraph was immediately confiscated by upper management and used in many high-level briefings. Imitation is the most sincere form of flattery.

It almost goes without saying that simply reading books and articles that reinforce what you already believe is not very productive. Your reading list should include topics that you don't necessarily agree with. True issues are rarely black or white, and you will be better off if you understand the arguments behind each viewpoint.

And finally, it doesn't hurt to read fiction. We believe that some science fiction, for example, should be mandatory for those who work in technology. This genre has inspired technical researchers in the past and helped create devices as varied as submarines and helicopters (Jules Verne), rockets and atomic energy (H.G. Wells), teleoperated arms (Robert Heinlein), and mobile phones (Gene

---

[27] Matt Ridley, *The Red Queen: Sex and the Evolution of Human Nature* (New York: Harper, 1993).

[28] Lewis Carroll, *Alice's Adventures in Wonderland and Through the Looking-Glass* (New York: Barnes & Noble, 2004), 175 (Chapter 6).

Rodenberry).[29] If nothing else, you can use examples from these books or movies to spice up a briefing. When we were interested in exoskeletons, for example, we extracted some great quotes from Robert Heinlein's *Starship Troopers* to help motivate audiences.[30]

---

[29] Mark Strauss, *Ten Inventions Inspired by Science Fiction,* http://www .smithsonianmag.com/science-nature/ten-inventions-inspired-by-science-fiction-128080674/ (accessed March 12, 2015).

[30] Robert Heinlein, *Starship Troopers* (New York: G.P. Putnam's Sons, 1959).

CHAPTER FOUR

Self-Deception

Falling prey to self-deception is one sure way to minimize your effectiveness in a failing-forward-fast environment.

Anyone who makes decisions for a living needs to understand self-deception. Quite simply, if you spend a lot of your time on issues that you believe are important (but aren't), or working with people you think are competent (but aren't), or pursuing technologies that seem real and timely (but aren't), then you are probably deceiving yourself.

Now you are probably saying: "I know folks who are like that, but that isn't me. My instincts are good." Maybe they are, but we doubt it. Everyone is subject to some degree of bias and self-deception, and the trick is to recognize this and learn to deal with it. Accept that bias exists. Accept that you are susceptible. Do your best to minimize its effects.

Cognitive Illusions

One of the first things to realize is that you have limits. Even your basic sensory perceptions - what you think you see and hear in your environment - are as much a reflection of you as any objective ground truth.

Scientists have developed a few tricks to help make some of these inherent limits more obvious to us. Did you know, for example, that you have a blind spot in each of your retinas? It lies at the point of the retina where the optical nerve is attached. You can't see anything in that portion of the retina, but your brain conveniently fills in the blanks. Don't believe it? Take a look at Figure 7. Hold the page about a foot from your eyes and move it slowly toward your face. Keep the right eye closed and focus on the right-hand dot with your left eye. At some distance, you will no longer be able to see the left-hand dot. There's your blind spot.

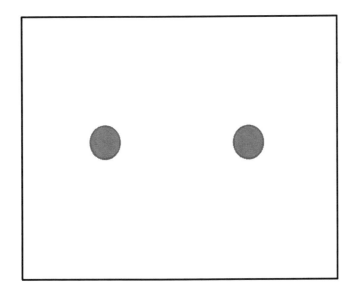

Figure 7. Finding Your Blind Spot

Another example can be found in a 1976 *Nature* article entitled, "Hearing Lips and Seeing Voices," in which authors Harry McGurk and John MacDonald describe how vision affects hearing.[31] The phenomenon has since become known as the "McGurk Effect," and you can see this on numerous internet videos simply by searching on that term. McGurk and MacDonald showed what happens when a soundtrack consisting of a series of pursed mouth sounds (*ba-ba-ba*) is synchronized with a video of an individual making a series of open-mouthed sounds (*ga-ga-ga*). Most viewers will hear something like *da-da-da*. When they close their eyes, however, they hear the *ba-ba-ba* soundtrack. The effect varies somewhat with language, but it is typically strongest when the sound is a bit corrupted. Somewhere along the line your brain decided that hearing is not just what your ear does.

While retinal blind spots and visual effects on hearing are fairly subtle, other deceptions border on the absurd. Consider, for example, inattentional blindness, or the problem of looking without seeing. Daniel Simons and Chris Chabris vividly demonstrated this effect in a classic experiment done at Harvard University and later published in *Perception* in 1999.[32] They asked subjects to watch a video tape of two teams playing a game. In this game, each team of three players passed an orange ball from one member of their team to another. One team was dressed in white shirts and one in black. One of the tasks was to count the number of times the ball was passed from one white-shirted individual to another.

The trick was that for about 5 seconds of the 75-second video, an actor dressed in a black gorilla suit wandered through the middle of the two teams while they were passing the balls. After the test, observers were asked a series of questions to see if they noticed anything unusual. More than half of the subjects never saw the

---

[31] Harry McGurk and John MacDonald, "Hearing Lips and Seeing Voices," *Nature* 264, (December 23, 1976): 746 - 748.

[32] Daniel Simons and Christopher Chabris, "Gorillas in Our Midst: Sustained Inattentional Blindness for Dynamic Events," *Perception 1999*, 28 (9): 1059 - 1074.

actor in the gorilla suit while they were focused on counting the ball passes. Their brains simply didn't think that the gorilla was important. This is an excellent example to bring up when someone claims to be very good at multi-tasking.

Cognitive Biases

The cognitive effects illustrated in the previous examples should alert us to the fact that we are not always aware of what influences us. We deceive ourselves when we think that we are completely objective and rational. The best that we can probably do is to try to understand what some of these biases are and adapt accordingly.

Many popular books and articles have been written on cognitive biases and why we make poor decisions.[33] Their authors catalog and describe many different types of biases and weaknesses. In our experience, four of the more important ones for program managers and COTRs to understand are confirmation bias, optimism bias, anchoring, and the distortions that accompany ownership.

Confirmation bias is the tendency to reinforce what you believe with more confirming information, while at the same time, discounting information that tends to refute it. No one escapes this curse. A failing-forward-fast environment depends heavily on your knowing when and how to change direction. If you ignore the negative signals, and only embrace signals that tend to confirm your present direction, then you will waste time and money that are better spent elsewhere.

---

[33] Some of our favorites that are easy reads include: *You are Not so Smart* by David McRaney; *Why We Make Mistakes* by Joseph Hallinan; *The Invisible Gorilla by* Christopher Chabris and Daniel Simons; and *Predictably Irrational* by Dan Ariely. *Thinking, Fast and Slow* by Daniel Kahneman is also a good read, but a bit more challenging.

67

Optimism bias is the tendency to overestimate the value of something and underestimate the associated risks. When we see this happening, it always calls to mind the *Far Side* cartoon in which two spiders are weaving a net at the bottom of a playground slide. The caption is "if we pull this off we will eat like kings."[34] Just like Gary Larson's spiders, it is easily to kid yourself into thinking that the value of your great idea (and the inevitable praise you will receive) is unlimited and that the risks are minimal and easily surmounted.

Anchoring is the idea that your decisions and beliefs can often be influenced by first perceptions. McRaney uses the example of an old sales trick in which the item you like is vastly overpriced, but just happens to be on sale the day you walk in the store. That first price tag ($1000) anchored you so that when you saw the sale price of $400 (probably still way too much), you whipped out a credit card without hesitation.[35]

We have seen anchoring effects in program managers that have been so strong that they forced individuals to quit or - even worse - give up in place. This can occur in government because of the variable nature of budgets. Once a program manager gets used to having a lot of funding for projects, it is hard for them to face lean times. Some of them can't shake the anchoring effect of the past and give up instead of adapting to the new conditions. It seems their egos get too tied up in how much money they control rather than the problems they are trying to solve. The smarter ones understand that funding is a cyclical variable, and that tough times are opportunities to get ready for the better times.

Finally, ownership distortion is the idea that we value our stuff more than others do. Many academic studies have been done that show this effect using examples ranging from scalped basketball

---

[34] Gary Larson, *The Far Side Gallery* (Kansas City: Andrews and McMeel, 1984), 140.

[35] David McRaney, *You are Not so Smart* (New York: Penguin Group, 2011), 214-219.

tickets to coffee cups.[36] For a COTR, the key ownership issue is the projects that you have invested considerable effort in. It can be hard to give them up when they are no longer valued by your customer, but you need to recognize that you might be valuing the effort more than an objective person would.

Self-Deception in Technical Work

Scientists and engineers are as susceptible to self-deception as anyone, but unlike most folks, they have the advantage of working in an environment that recognizes those weaknesses to some degree and takes measures to counteract them. Many of the tools of science such as double-blind studies and peer review are designed to minimize the effects of self-deception.

A good description of how self-deception can affect scientific work is described in Irving Langmuir's classic seminar *Pathological Science.*[37] In this 1953 talk, Langmuir discusses what he terms, "the science of things that aren't so," and the pitfalls that can put scientists into these predicaments. He describes examples of bad science and provides an analysis of their common traits.

Some of the characteristics of pathological science involve the nature of the measurement, such as when the effects being measured tend to be close to the limit of detectability for the instrumentation being used. But others deal with the behavior of the individuals who are collecting and analyzing the data. Theories that are contrary to experience, ad hoc responses to criticisms of the work, and the dismissal of negative data and

---

[36] See, for example, Daniel Ariely, *Predictably Irrational: The Hidden Forces that Shape our Decisions* (New York: HarperCollins, 2008), 127-138.

[37] Irving Langmuir, "Pathological Science'" *Physics Today* (October 1989): 36-48. Langmuir was a chemist who received the Nobel Prize in 1932 for work in surface chemistry.

embracing of positive data all suggest bias and thus potential problems with results.

One of the best examples of pathological science that Langmuir describes is the investigation of N-rays. This story involves two men. One was Robert Wood, a professor of physics from Johns Hopkins in the U.S. The other was Rene Blondlot, a professor of physics at the University of Nancy in France. Both were well-respected scientists.

In 1903, Blondlot announced that he had discovered a new form of radiation which he called N-rays. This radiation could be produced by a variety of materials, and prisms could be used to refract and separate the N-rays by wavelength. In one experiment, for example, Blondlot produced the rays by heating a ceramic rod to incandescence, filtering the non-N-ray radiation, passing the filtered result through a slit in an absorbing material, and then impinging the N-rays on aluminum prism. A phosphor was used to detect the resulting spectrum.

Blondlot published a number of papers on various aspects of N-rays, and once the discovery became well known, other scientists began to publish articles on the subject. But there were problems. In particular, some well-known scientists were having difficulty duplicating Blondlot's results. At the urging of some of his colleagues, Wood visited Blondlot's laboratory in order to observe the measurement process first hand.

During this visit, Wood was particularly interested in the optical design of the experiment described earlier, and why Blondlot was able to measure wavelength so accurately when ordinary physics says he shouldn't be able to.[38] Blondlot's response was: "that is

---

[38] Among his many concerns, Wood did not understand how a beam of radiation coming from a broad slit could be resolved so finely into the components that Blondlot claimed.

one of the fascinating things about the N-rays, they don't follow the ordinary laws of science."[39]

A second problem was that Wood could not see the effects that Blondlot claimed. Wood then got a bit sneaky and tricked Blondlot into doing a series of experiments in which the aluminum prism was unknowingly removed. This change in configuration made no difference to the results.[40] Wood published his observations in *Nature*, and that pretty well settled the matter, except for Blondlot, who continued to believe that N-rays were real.

The CIA is no stranger to pathological science. Much has been written, for example, about remote viewing, a branch of parapsychology. Remote viewing is the idea that someone can mentally perceive accurate details about an inaccessible target. Say, for example, country A has built an underground facility in a remote area. Other countries, including the United States, will want to know the purpose of that facility as well as many of its pertinent details. Clearly, remote viewing would be one low-risk and low-cost way to acquire such information, if it was real. Unfortunately, it does not appear to be effective. As Carl Sagan was fond of saying, "extraordinary claims demand extraordinary proof."[41] Remote viewing never met that standard.

The espionage business seems to attract people with extraordinary claims, and investigating them was just part of the job. For example, we once visited a midwestern contractor who claimed to have a device that could monitor the mental state of an individual or group of individuals using a simple remote radio-frequency technology. After listening to the contractor's arguments for

---

[39] This is a big warning flag. We have heard similar statements on more than one occasion.

[40] The stories told about this visit vary a bit in the details. Our account is largely taken from Langmuir's seminar and Irving M. Klotz, "The N-Ray Affair," *Scientific American* (May 1980): 168-75.

[41] This quote has been attributed to others as well, including Pierre Simon LaPlace.

several hours and asking many questions, we were asked what we thought of the work. We told them about Langmuir and pathological science, and systematically went through some of its characteristics and how they were similar to what we just heard. None of these individuals that we met with was the least bit shady or misleading. They (like Blondlot) were sincere, well educated, and successful in their own fields. They wanted to help. While our explanation did not sway them, it did solve our immediate problem which was whether or not there was a next step for us to take with this technology.

There is an endless source of stories about individuals and groups of people who deceived themselves in some manner or another. Perhaps the best book on the subject is Charles MacKay's *Extraordinary Popular Delusions and the Madness of Crowds*.[42] Although published in 1841 and focused on self-deception in earlier times such as the 17th century Dutch tulip mania, it is still relevant. The chapters on economic bubbles are particularly insightful for their resemblance to the dot-com and real estate fiascos that occurred at the beginning of the 21st century.

Peer Review: Your Best Weapon against Self-Deception

Whether you are a scientist or not, peer review is probably your single best weapon against self-deception. This process can be as complicated as working a paper through a journal's peer review and publication process or as simple as talking through some issue with people you trust.

This process is not foolproof, of course. Even scientific peer review does not always get things right, despite blind review and scrutiny by editors. An editor's purpose in peer review is to decide

---

[42] Charles MacKay, *Extraordinary Popular Delusions and the Madness of Crowds* (London: Richard Bentley, 1841).

whether or not to publish something in its current form or some modified form. The most important aspect of peer review, however, is not the journal's decision to publish, but the exposure of the paper to a larger audience. Once the paper is published, others see the work and sometimes try to replicate the results. Errors are found, fraud is sometimes detected, and good findings are built upon.[43]

Holding your work up to criticism can be hard, but it is essential for avoiding self-deception and coming up with the right next move. Unfortunately, some people take any form of criticism personally. If you are one of them, try to get over it or find another business to work in. That person who took the time to put together a thoughtful criticism of your work is doing you a big favor, even if you might not believe it when you get the critique.

Neither of us started out with the thick skins we now wear. It was only over time and with repeated reviews of our data and conclusions that we really began to understand how valuable criticism can be. Compare, for example, where you start out with a pitch or an idea or an analysis and where you end up after you have adapted to the many criticisms and comments that you have received about it.

Unfortunately, not everyone who gives criticism is good at peer review. There are many in government who think that they are God's gift to the organization and relish the chance to put down your work. The criticism is not intended to help you, but rather to feed their own egos. Stay away from these people if you can and - if you can't - work hard to defend your work or position. You might never convince them, but others who are observing probably understand what is going on and will credit you with a win.

---

[43] Jordan Ellenberg refers to replication of results as a sort of imperfect immune system that rids us of results that do not belong. See Jordan Ellenberg, *How Not To Be Wrong* (New York, Penguin Press, 2014), 161.

There are good ways and bad ways to give criticism. Good relationships can stand almost any approach, but under more formal circumstances like a briefing, it is best to temper your words. Questions are less threatening than judgments or blanket statements. For example, try using questions like: "Have you ever considered...," or "What would happen if...," rather than emphatic statements like: "That can't be right," or "That will never work," or (our particular favorite) "We tried that a long time ago."

It is always easy to say that something won't work, particularly in fields like espionage where the problems are really hard, have been studied by many, and are fraught with ambiguity. But just tearing down a proposal isn't constructive. A better approach is to ask yourself what can be done to make the proposal work and tailor your comments appropriately. As Robert D Gilbreath noted in his book, *Escape from Management Hell*: "Innovation is never executed, it is pecked to death by chickens."[44] Don't be one of the flock. Government already has enough chickens.

Keeping an Open Mind

In a failing-forward-fast environment, you must be able to change your mind in the face of reasonable evidence. If you don't, everyone suffers. You pay for the mistake by wasting even more of your time on the way to a solution. Your customer pays because he or she doesn't get what was expected. Your contractors waste a portion of their careers on things that won't pay off as intended. And the taxpayer loses because you keep dumping their money down your newly-discovered rat hole. If you fall hopelessly in love with a particular approach to a problem, or you have so much invested in it that you cannot bear to abandon it, or you have been overselling and can't bear to swallow your pride, then you are not moving forward.

---

[44] Robert D. Gilbreath, *Escape from Management Hell: 12 Tales of Horror, Humor, and Heroism* (San Francisco, Berrett-Koehler Publishers, 1993), 39.

Many government efforts go much further than they should before being cancelled or altered significantly. For a COTR, the key question is whether you should continue to do something based upon the new information you receive. Before a review you have some level of confidence or belief that a project will be successful. After a review that shows positive results and progress, you probably have a higher level of confidence. But if the review is negative, your confidence in the program might go down.

Unfortunately, it is not always that simple. Your impressions of that review could be right or wrong. People have bad days and good days. Quick answers to questions might or might not be the final word. New test results might be right or wrong. The test might or might not have measured what you think.

You really earn your pay at such times. Dig in, review the results, and make sure you understand what was really done. Show the same degree of skepticism for positive results that you are likely to show for negative results. Repeat the test or refine the test as needed. If it makes sense, do the review over in a couple of weeks. If nothing improves, then seriously question whether the effort should proceed as planned. You don't want to brush off new negative data or blindly accept positive data because of unrealistic confidence in your current path.

Staying in Touch with Reality

You will occasionally find yourself visiting contractors and being dissatisfied with progress. Sometimes this is indicative of a problem with the contractor. Perhaps they really weren't working on your project until you called and said you were showing up next week. Other times the trials and tribulations of life are getting in the way. Contractors get sick, contractors have bad weeks, and contractors make mistakes.

But it is also possible that you are part of the problem because you are losing touch with reality. We first noticed this effect as graduate students at Penn State. One of our professors started to get impatient with our rate of progress on a study. Not used to being called slackers, we started to look a bit more deeply and realized that the problem was the professor's unrealistic expectations. He had been away from the lab too long and forgot that everything doesn't go as planned. This was a case of a manager thinking that every nail always goes in straight, every board is cut to perfect length, and every part shows up when expected. Things are always simpler when you only have to plan something new, rather than actually do it. Sometimes your calibration point slips.

One of the best ways that we know to minimize these effects is to do some research yourself. We can already hear some of you saying "Are you crazy? I don't have enough time already!" But if you see yourself slipping, this might be the simplest and most direct way to recalibrate. Even former Secretary of Energy Stephen Chu continued to do research during his tenure at DOE. Some of you might also say that you don't have the equipment to do research to test ideas and assumptions. While we believe everyone should own some basic test equipment if for no other reason than to solve home problems, we know that it is hard to justify a universal testing machine or other specialized tools that your wife or husband might not like seeing in the living room. Fortunately, you have other options. Just put a task in the contract that provides you with laboratory access somewhere. You might be surprised at how your perspective changes when you have to do lab work yourself again.

CHAPTER FIVE

Failing-Forward-Fast Program Management

Key job elements include: finding good customers and contractors, getting the requirements right, knowing when to stop something, and figuring out whether the time is right for a particular technology.

In the Agency, we were often frustrated by how slowly the paperwork got done. When we criticized Agency managers about this failing, their response often was: "Yeah, we're a bureaucracy, but when we have to, we can move fast." While we are sure that the comment was meant as a source of pride to them, it amazed us that they missed the insult they were conveying to everyone: "Oh, yeah, we can do this faster, but your time isn't important enough to do so."

The slow pace of government might not matter to a bureaucrat, but it should matter to you as career government scientist or engineer. You have only so much time and energy to make a difference. If you are young and just getting started this might not seem so important to you. But the point was driven hard into us late in our careers when we made a comparison between our early passport photos and later ones. We flipped a coin and Greg lost, so Figure 8 shows what time did to him. Not a pretty sight, but it really drove home to us that there is no do-over button for a career.

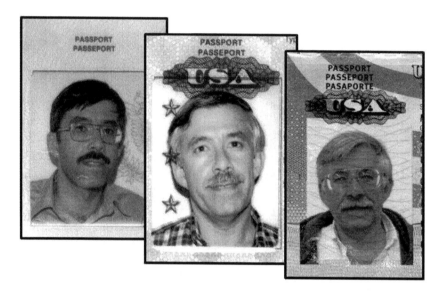

Figure 8. Early and Late Passport Photographs at the Agency

As a program manager, you should not allow the government and its processes to marginalize your efforts any more than necessary. Government will never be a race car, but you can have some control over the throttle for your own projects. Here are some guidelines that helped us.

Keep Management in the Loop

Management support is essential for success in a failing-forward-fast environment. This means both resources and the right amount of help.[45] This support provides the foundation that you need to build your projects. Just as Newton attributed his success to "standing on the shoulders of giants," you need to plant your heels on the shoulders of management. They have to nod "yes" or you aren't going anywhere. Going on your own is suicide and can hurt others as well.

Keep in mind that you don't always have to be the one lobbying your management for the attention, approval, and resources that are required. Anyone who has the ear of management can also help you to get the support you need. Many of our own conversations with upper management, for example, involved them asking what we thought of someone else's work, rather than anything that we were managing. By informing key individuals of your thoughts and efforts, you are in effect expanding your virtual organization.

How do you gain such support from others? Part of the answer is by doing a good job in reviews where they are present. It also helps to discuss the work directly with them so that they better understand the ideas and challenges and believe that you have covered the relevant issues. Getting a bit of help from others also subtly increases their buy-in. But the larger part is probably the image you convey of yourself, such as your thoroughness and willingness to ask and answer questions.

Be Suspicious of Requirements

When you are doing research in a failing-forward-fast environment, it helps to think about requirements a little bit differently from most procurement situations. Rather than thinking of requirements as absolutes, think of them as negotiable project

---

[45] Too much management help, of course, is counterproductive. As one of our colleagues was fond of saying: "pray that you do not receive as much help from management as you are paying for in overhead."

parameters. While this sounds a bit heretical (shouldn't everyone in government use the same definition of requirement?), it takes advantage of the fact that as a project progresses, the problem and the solution both become clearer. In an operationally-driven environment, we liked to think of a requirement as something that has three interrelated attributes: need, means, and priority.

A need is what the customer says he or she wants. It is a starting point that will ultimately be modified by the limitations of technology, changing operational constraints, and priorities. Needs are often overstated at the beginning and require solutions that are far more difficult technically than they need to be. Sometimes needs don't even address the real problem. But if you can ask the right questions early on and throughout the project, both you and the customer will gain a better level of understanding, and that often leads to a solvable problem. The time you spend clarifying needs is not wasted. Any insight that tells you that you are on the wrong path initially might seem like a setback, but it is a critical part of failing forward. Catch these mistakes early when you can because the longer you are on the wrong path, the more expensive it becomes.

The second element is means. What are the solutions that might meet the customer's needs? Don't wait for a competition to start discussing solutions. One of our favorite people used to remind us often that "hope is your enemy." If you don't have plausible approaches and are hoping that some sort of competition will provide them, you run the risk of getting nothing that will work. You also run the risk of not meeting the true need that would have emerged through earlier discussion of possible solutions and needs before the RFP went out. Showing a customer potential solutions is a good way to elicit details and constraints concerning their needs that were not raised earlier. Clearly, it is better to have at least some of these trade-off discussions with the customer before the contractor spends bid and proposal funds.

The third element of a requirement is priority. How important is the problem now, and how important is it likely to be two years from now when you have a final solution at hand? A crisis that has the organization turned upside down at one moment can be viewed with a yawn as "yesterday's problem" in a few years. Or a problem that was heartily endorsed by customer management when you started working on it could later be discounted by new management.

A second way to think about priority is the importance of some parts of your solution compared with other parts. If you really have an important problem to solve, and you can get 90% of it easily and with low risk, you really need to know how important that last 10% is. Your customer needs to know as well because chances are that if you toss or moderate that last 10%, you might be able to turn an impossible problem into one that is readily solved. Not all aspects of a problem are equally important to a customer, and it is important to find out which ones really matter and whether workarounds are possible with the others. Such things become much clearer when the schedule is getting short. One of our favorite DCIs, Bill Casey, was fond of the saying: "Better is the enemy of good enough."[46]

Put the three elements together and you have what we consider a valid requirement. First is a need that must be met. Second is a means or approach to solve it. Third is the priority that characterizes the need and the method.

We admit that our approach does not neatly fit all cases. In our world, a tradeoff existed between what a human asset could do and what technology could do. If the technology was too hard, then perhaps an operational change was possible. Too often, people wanted technology to do the whole job, and that sometimes led to ridiculous requirements that chewed up time and resources. Good customers understood this reality and worked with us to find a solution.

---

[46] The quote is not a Casey original. Variants have been attributed to Voltaire and Von Clausewitz, among others.

At times, you will be tempted to accept the requirements that a customer provides initially because it is easier in the short term, or meets some schedule requirement. But as we have said, requirements can be wrong and are frequently bloated with unnecessary restrictions and uneven priorities. To accept them simply wastes precious time and resources. Spend the time up front to get as much right as you can, and be sure to build sufficient flexibility into your plans and contracts to deal with the remaining uncertainties. Simple, focused experiments done before you commit to a large contract can often provide you and the customer with useful insight and better questions.

Finally, the process is almost always better when the customer has some skin in the game since you are both spending money on a common goal. If the customer is only providing requirements, they have less at stake and might be less willing to compromise.

Find Multiple Reasons for Doing Projects

Circumstances can change quickly in an operational world, and what was relevant yesterday might be irrelevant today. One of the best ways we found for avoiding this "yesterday's news" trap was doing projects for more than one reason. If all or part of the solution you are working on can be applied to multiple operations, then the chances are better that it will be adopted. In this respect, recurring problems can be more productive than unique ones. If the current mission requirement is dropped while you are focusing on a recurring operational problem, you still have some reason to continue. It is a bit like buying an insurance policy for your project. The problem will probably show up again, and you will have a ready answer.

Negotiate with Customers

"The customer is always right" is not is a good rule to follow. You need to negotiate with your customers if the organization is to truly benefit from your efforts. Here is one example.

The CIA must to be able to create new identities for people so that they can move about the world undetected. As a result, we spent some of our early years building tools and techniques that helped to address the Agency's needs for disguises and false documentation. A good declassified example of this operational challenge was described by Tony Mendez in his book, *ARGO*, and in the subsequent movie of the same name.[47] The story involves an operation that occurred during the Carter Administration. On November 4, 1979 a group of Iranian revolutionaries overran our embassy in Iran and captured most, but not all, of the employees. With the help of the Canadians, six U.S. diplomats successfully hid from the insurgents, and the Agency's job was to get these hidden people safely out of Iran. Tony and another Agency officer accomplished this using various disguise, forgery, and tradecraft techniques. This was no remote border crossing on a moonless night using night vision goggles. Tony and his colleague exfiltrated the diplomats (disguised as members of a movie production company) through the Tehran airport, right under the noses of the insurgents.

Doing research for an operational support group such as Tony's was challenging because most of their energy was understandably focused on the current crises. Getting agreement over issues that directly impacted current missions was easy, but longer-term issues could be tough. For example, a new way to do color matching might meet with "we have always done it this other way," even though a stronger scientific basis could ultimately

---

[47] Antonio Mendez and Matt Baglio, *Argo, How the CIA and Hollywood Pulled Off the Most Audacious Rescue in History* (Viking: New York, 2012).

result in a more robust product. It was the old challenge of being too busy sawing a board to take time to sharpen the saw.

How did we get at those longer-term issues while also addressing the customer's top priorities? We simply proposed an 80/20 solution. Eighty percent of the budget would go to traditional priorities and 20% would go toward those longer-term capabilities. In this way, everyone was fairly happy and we were able to provide some long-term and short-term balance in the research portfolio.

Find and Keep Good Customers

You will be more effective if you spend time with only good customers. We really upset our group chief once by suggesting that we should not do any additional work for a particular development group because they had a poor track record of integrating technologies that they had not invented themselves. We had a good reason for stirring this particular pot. We saw a number of very good young COTRs being tossed into this meat grinder, working really hard, and always losing. In the end, they were not making a difference. Unfortunately, we lost that battle, and could only help by advising COTRs to avoid that particular customer

Here are some signs of a good customer:

- They are willing to inconvenience themselves for you.
- They are willing to argue with you in a productive way.
- They give you credit where credit is due.
- They readily acknowledge (at least privately) when you take a hit for them.
- They do what they promise to do.
- They give you feedback on what happens.

- They are honest about what they know and what they don't know.
- They invite you to meetings about problems and new opportunities.
- They are in tune with organizational and operational realities.

A bad customer, on the other hand:

- Doesn't keep you informed.
- Has management that doesn't care about the effort or perhaps doesn't even know of the effort.
- Wants to run the effort.
- Creates unnecessary problems for the contractor.

When you find a good customer, try to stick with them. Even if they change jobs, consider redirecting your research and continuing to support that customer in his or her new role. A good customer is really hard to find, and the trust that is built up over the years between two groups of people is priceless.

Sticking with the same customer is one key to involvement in many different projects and operations. For example, John was one of our best customers. When we first met him, he was on the operational side and needed a technological solution to a particular problem. We came up with an idea and built several successful systems. Eventually he moved to another job, where we helped him develop and successfully deploy another new capability. Several years later he moved again and we continued to support him in his new area of responsibility. John was essentially part of our virtual division or splinter group and we were, in turn, a part of his. We were able to mesh our talents and share many successes when opportunities arose.

Learn When to Stop

It is human nature to preserve something that you have spent a great deal of time and money creating (remember the ownership distortion in Chapter 4?). You get emotionally tied to the work as you invest more and more in it. Most readers can probably recall a time in their career when a government program manager admitted that he or she did not want to hear about other options since the program was "too far along." It is one thing if you are in the last phase of a long project and about to receive deliverables. But this mindset can also creep into the process much earlier - even into the planning stage. You might not want to go back to management with: "Belay that last briefing. There is a better way." But to fail forward fast you really need to be able to redirect or stop yourself and your contractors whenever it is in the customer's best interest.

Perhaps the thought of changing or even killing your own project makes you sick to your stomach. After all, you have a lot invested in it. But such decisions are part of your job as a program manager. You need to plan your contracts to accommodate surprises and new information. Focus the contractor on what you believe will work and develop options in case things do not go well.

It might help to think about pulling the plug on something as a selfish act. The less time you spend on an idea that won't pay off, the more time you have to work on things that will. The same is true for your team. But what about the funds that have already been spent? These are sunk costs that shouldn't drive any decision to stop the work. The Defense Advanced Research and Projects Agency used this philosophy and did not base a decision to continue on what was already spent on the project. They understood that the real issue is limiting non-productive future investment. We think that is the right approach, but also recognize that it is not easy to pull the plug on something that you and others have worked hard at.

Investigate Multiple Solutions

With operational problems, we often had more possible solutions than the available funds would cover. Early on in the process, some of those potential answers looked better to us than others, but we were often guessing. The right answer was not to immediately down-select to one that we then took to completion, but rather to analyze, experiment, and test as many as we could, and use the resulting information to narrow the choices.

For example, we recall one problem for which we had identified over a dozen different approaches. There was no existing solution at the time, and no way to fund all of those ideas to completion. So how did we fit this into the available funding? First, we winnowed the number down to something more reasonable by eliminating as much as possible based on what we knew or learned in the short term. We also integrated some options into hybrid approaches, since there was evidence that the combinations might yield better results than each approach alone. In the end, we were left with two basic approaches and two hybrid techniques. Even with that reduced number, however, there was still not enough money to run all to completion.

We needed a contractual mechanism that would allow us to quickly examine the options in order to tell whether or not each was a fruitful approach. We needed to be able to go down a particular path, see what happened, get out of it if didn't meet our needs, and move on to another approach. To acquire this flexibility within one contract we went to a senior person in the contracting staff and proposed a contract in which we could verbally change the contract direction at the contractor's monthly review meetings. Once the contracting officer understood our dilemma and what we were trying to accomplish, he agreed to the request as long as there was an adequate audit trail to track spending.

The contractor doing the work was used to taking weeks in order to make even small changes to a contract, so this new approach represented a big shift in their thinking. To ease them into this new process, our senior contracting officer visited the contractor with us to explain how the effort would be run and to answer any questions. In the end, the contractor agreed that this was worth trying in order to maximize our output for the limited dollars that were available.

While we now had this mechanism for moving quickly, we knew that we also needed as much help as we could get in order to know when to pull the plug on an approach. In the same meeting we came up with the idea of giving an award to any engineer who killed his or her project so that the money could be better spent elsewhere. The contractor decided to call this the "Outer Limits" award after a 1960s television show that focused on fantasy and supernatural phenomena. This was later briefed to the contractor's senior management, and they were sufficiently intrigued with the idea that they offered to pay for the award for their people.

We added the Outer Limits award to the process because we knew that those closest to the experiment are sometimes first to sense looming disaster. The award provided some incentive for the engineer doing the work to end the misery rather than ride it to the end.

This award program turned out very well. It helped us to examine key issues more efficiently and quickly move on. Government contracting is not known for speed and efficiency, but if you put the right people together and clearly explain what you're trying to do (and that you'll take the blame for failure), you can sometimes pull such things off.

Form a Personal Advisory Group

In a failing-forward-fast environment, the right kind of advisory group can make a big difference. Ours was called the Gang of Six. We started a group by having each of us select three people we had worked with in the past and considered to be highly practical. Each had built many things and was a known and respected problem solver.

The Gang of Six was an important part of our failing-forward-fast environment because the team would not only produce new ideas, but would also develop and test the ideas quickly for minimal cost. Our boss was always surprised to find out how many efforts we were tracking (it was in the hundreds) in order to decide what new projects to start in the future. The Gang of Six was one reason why we could maintain such high leverage.

As problems changed, we formed and disbanded three different Gangs of Six. Most of our candidates for these groups had not worked together before, and our biggest challenge was to make sure everyone trusted us to police the team, if needed. Once everyone felt comfortable, things went remarkably well. We have rarely seen a comparable degree of peer review, and it is fair to say that those meetings were some of our best days in government.

Be Wary of New Research Results

Most researchers would agree that spending some time studying research magazines is a good practice. We particularly like *Science* and *Nature*, and you probably have your own favorites. Such activities can be hazardous to a program manager's health, however, since it is easy to get caught up in the excitement and the hype that new research results often engender.

One of the research areas that got our attention was gecko adhesion. Geckos are lizards that have a remarkable ability to climb a wide variety of vertical surfaces. To a program manager in defense or intelligence work, the idea of mimicking this movement in climbing robots was enticing.

We were pretty excited, therefore, when scientists at the University of California at Berkeley published an article entitled "Adhesive Force of a Single Gecko Foot-Hair," in a 2000 issue of *Nature*.[48] In it, the authors showed that the fine hairs on the surface of the gecko foot could generate significant attachment forces to various surfaces when presented in the proper way.

The Berkeley work inspired additional research that included further investigation of the adhesion mechanism as well as efforts to produce adhesives based on the gecko structure. Berkeley kept a nice bibliography on the open literature related to gecko adhesion, showing that in the years 2000 to 2012, over 350 papers were published, representing contributions from about 550 scientists.[49]

So where do we stand today, fifteen years after the publication of the first *Nature* article? To our knowledge, there are no successful commercial or government-fielded applications of this technology. No adhesive bandages at Walgreens. No adhesives at Best Buy for attaching flat screen TVs to walls. No operational military robots that climb like geckos. No gecko tethers for aircraft on carriers.

But gecko adhesion is far from a dead end. Along the way, for example, researchers at NASA's Jet Propulsion Laboratory demonstrated that the technology could be used to grab space

---

[48] Kellar Autumn, Y. Liang, W. Zesch, W.-P. Chang, T. Kenny, R .Fearing, and R.J. Full, "Adhesive Force of a Single Gecko Foot Hair," *Nature*, 405, (June 8, 2000): 681-685.
[49] Ronald Fearing, *Gecko Adhesion Bibliography*, http://robotics.eecs.berkeley.e du/ ~ronF/Gecko/gecko-biblio.html (accessed January 18, 2014).

debris.[50]   On the military side, the Defense Advanced Research
Projects Agency (DARPA) started the Z-Man program to use these
technologies to develop a way for soldiers to climb surfaces
without the benefit of a ladder. In 2014, DARPA provided an
impressive demonstration of this application when a 218-pound
man carrying a 50-pound pack successfully climbed up and down a
25-foot high glass wall using gecko-inspired paddles. According
to the news report on the 2014 test: "The paddles are not
battlefield-ready yet and tests of the new technology are still
ongoing."[51]

Since fifteen years can easily be half of a career in government, the
message should be clear - be careful with new discoveries. They
are not always what they seem to be at first, and require some
seasoning before the practical results appear. If your organization
rewards you for funding seminal peer-reviewed papers, you should
perhaps jump in with both feet. Just make sure the government
gets unlimited rights to whatever is found. But if you are a
program manager whose job is to put things in the field, you might
want to wait a bit and let others spend their funds working through
the basics. Follow the research, ask questions, and make
suggestions; but wait for the right time to jump in financially.

In the interest of full disclosure, we did spend some of your hard-
earned tax dollars on gecko adhesion research. But we didn't do it
because we thought it would solve some pressing field problem in
a timely fashion. We did it primarily because it helped
management achieve another goal.   Around the time that
researchers began to understand the nature of gecko adhesion, the
Intelligence Community started a postdoctoral research program to
fund basic and applied research at universities and national

---

[50] NASA Jet Propulsion Lab News, *Gecko Grippers Get a Microgravity Test
Flight*,   http://www.jpl.nasa.gov/news/news.php?feature=4421   (accessed
September 20, 2015).
[51] Kelly Dickerson, *Gecko-Inspired Tech Lets Humans Scale Walls Like
Spiderman,* http://www.foxnews.com/science/2014/06/10/gecko-inspired-techlet
s -humans-scale-walls-like-spider-man/ (accessed September 20, 2015).

laboratories. We (as program managers) were expected to support that program with some of our resources. The program was a good corporate idea for a number of reasons, but it was hard to implement in a largely classified world. Gecko adhesion, however, was one of the things we cared about that was suitable for the academic world of postdoc research. While none of the work led directly to a useful tool for us while we were there, everyone (including Congress) loved the geckos.

In the aftermath of 9/11, we shared an office for a short time with a former special operations officer who had retired and moved on to government contracting. He was bragging one day about how quickly he was able to turn some technology into a fielded product. Once he was done patting himself on the back, we asked him to do us a favor. The next time he decided to tell that story, why not give some credit to the researchers who spent 15 years getting the technology to the point where he could actually do something with it? The look on his face showed that he clearly had never given that any thought.[52]

Ask the Most Important Question

One of the most important skills that a program manager can possess is the ability to ask questions. It is the most effective way to learn and to get at the real issues before they blow up. Entire books are devoted to this subject, so we are not going to rehash techniques and strategies here.[53] But we will offer what we think is the most important question, and it is this: once a decision is made, ask why it is wrong.

---

[52] There is a variant of a Chinese proverb that goes: when drinking the water, don't forget those who dug the well.

[53] See, for example, Dorothy Leeds, *Smart Questions: The Essential Strategy for Successful Managers* (New York: Berkley Books, 2000).

For example, we recall a project where the first order of business was to evaluate different concepts for a covert communications system. The contractor had prepared a half-day presentation on why a particular option was the way to go. We immediately reversed the situation and told them we accepted their results but now wanted them to tell us every reason why we'd be kicking ourselves in 12 months for going ahead with this option. This threw the group for a loop, so we started calling on people to give us all of the negatives they could think of. Finally, one person pointed out that the recommended option could come close to one critical requirement (a detection issue) but never meet it. This realization caused an immediate reversal of the contractor's opinion on which option was best.

Do not be fooled into thinking that you are doing the right thing just because people seem to be in agreement. As this example shows, the desire for conformity can sometimes give you the wrong answer. Part of your job as a COTR and program manager is to make sure that people do not suppress important issues in the rush to agreement.

Don't Ignore Problems

In general, you want to fix problems when they arise. Waiting might help in some cases, but many problems require immediate attention or they get worse.

One of the more difficult problems that you will deal with as a program manager is an ineffective program manager on the contractor side. Some are incompetent, some think that their way is the only way, some are deceptive, and some have personalities that just don't work with yours. It is not easy to go to a contractor's senior management and request that a person be removed, but both of us have had to face this problem. Not all contractor PMs can deal with a failing-forward-fast environment and you need to avoid them.

In the few cases where we did not request the change and gave the program manager an opportunity to change his behavior, we were disappointed and the work suffered. In one case, a chip-design firm replaced our program manager with an individual that we did not know. As we tried to understand this sudden change, we talked with the former program manager. He told us to be wary of the new person because of his background. Essentially, he would focus too much on one aspect of the problem that was more meaningful to him than necessary for the project. We did not know at the time whether this comment was sour grapes or a real issue. We met with the new program manager specifically to explain to him where we had to be when the money ran out. We emphasized that ending the phase without some clear view of whether this idea was going to work would be the worst possible outcome. Despite this new PM's assurances, we ended up exactly where we didn't want to be. We ended the program at that point since we no longer trusted the company or its management.

No matter what problem you are solving, make sure that the problem is serious and not simply something being blown out of proportion. It is easy for people to get worked up over nothing. An innocent mistake like a minor security violation, for example, is no reason for a hanging. On the other hand, problems like discrimination, abuse, and fraud are big deals. On average, experienced officers tend to have a more seasoned view of the world and understand what is a big deal and what isn't. If you are not sure, talk with some of them first before creating a fuss that might not be productive.

Learn to Deal with Sudden Fame

When you become a government program manager or COTR, you will suddenly be very popular. Folks who normally wouldn't give you the time of day are now pushing their business cards at you. When this starts to happen, repeat after us: "It's not about you, it's about the money." We first saw this saying on a DARPA office director's white board, and think it is some of the best advice that any government program manager or COTR can get.

Being popular creates problems and opportunities. One problem is that some folks will start to kiss your rear end in the hope that if you like them, you will pay them. You do not need that type of support. Another problem is that you are now a gatekeeper for a lot of valuable information, and some folks will try to elicit that information from you.[54] You must therefore always keep your guard up. For all of this pain, you do get something good in return. More ideas will come your way as companies and universities and national laboratories see you as a possible source of funding. You might have to sort a lot of chaff to find the wheat, but it is probably easier than trying to find out what folks were doing when you were nobody.

You should not feel guilty about this sudden fame. Instead, enjoy the access that it conveys, because it will not last ("It's not about you..."). We have met some really great people this way.

---

[54] A good book on business elicitation is John Nolan's *Confidential: Uncover your Competitors' Top Business Secrets Legally and Quickly --and Protect Your Own* (Dunmore, PA: Harper Business, 1999).

The Bottom Line on Program Management

We have known many COTRs and program managers who thought that managing programs was like following a recipe. Sadly, they believed that everything would turn out fine if they just followed "the process." But being an effective government program manager requires much more from you than just blindly following the procedures that you learned in government courses. Don't let the bureaucracy turn you into a management-and-contracts automaton.

CHAPTER SIX

Contracts, Legal, and Unwritten Rules

Getting the right contract in a timely manner is critical if you want to fail forward fast.

If you are like most industrial or academic scientists and engineers who enter into government, you probably began with very little knowledge of government contracting. Some of you might have even been a little snobbish about it. After all, weren't you hired for your technical expertise? Doesn't someone else handle "the paperwork?"

The Agency maintains a great deal of expertise in specialized areas such as the tradecraft for using equipment operationally, but the development and production of that equipment is often done by contractors. The reason is that the pace of technology and the costs of recapitalization far exceed the ability of government to keep up with them. Imagine, for example, the challenge of trying to maintain all of the required research, development, manufacturing, and maintenance capabilities in fields like communications, materials, sensors, computation, optics, and image processing, to name a few.

While you might be disappointed at first that you do not have a personal state-of-the-art laboratory to hide in all day, eventually you will realize that you have much more. With the right contracts, the best labs and manufacturing facilities in the country can be at your disposal. One of your first goals after joining government as a scientist or engineer, therefore, should be to become a COTR.

Becoming a COTR

Government does a pretty good job of providing COTRs with the basics of contracting – things like terminology, tools, processes, etc. This material is often boring, but you have to know it. When you finally get the chance to admire that framed government-issued certificate of completion of basic training, remember that nine-out-of-ten experienced COTRs and program managers agree that there is nothing so dangerous (or obnoxious) as a brand new COTR who has just completed training and has no real experience.[55] That certificate is a learner's permit and nothing more.

As you do more and learn more, you will begin to realize that the tool set provided by the government is not sufficient to do a good job. Other skills such as those we mentioned in Chapter 3 are important too.

Do not expect your organization to provide you with these other skills as part of your ongoing COTR training. And don't depend on your annual review to provide much insight, either. Your development is mainly your problem, and there are many ways to

---

[55] We never really took a poll on this, but almost everyone we know would agree.

get the skills you need. In our case, management was pretty good about allowing us to take the courses we thought we needed and to work for extended periods at contractor facilities. If you don't know what you need, ask those whom you trust and respect for advice.

You will probably not agree with the entire curriculum that the government considers important. In our case, required COTR training was largely determined by senior contracting officers rather than senior COTRs, which meant that some aspects of the job (like contracting) were overemphasized and other critical skills (pick almost any other chapter in this book) were largely ignored.

Another problem with government training was its one-size-fits-all philosophy. This meant that COTRs were often required to take training that had little relevance to their job. At one point, for example, all COTRS in the Agency were required to take additional training in the legal aspects of contracts because of several large contract overruns that had occurred. After some digging, we found out that these contract overruns occurred exclusively in our services directorate that built new buildings and controlled the Agency's property and utilities. Several senior managers in our organization saw right through this exercise and waived the training for all of their COTRs who had nothing to do with construction and maintenance projects. Most Agency managers meekly went along with the decree, however.

If you are stuck taking something that seems like a waste of time just grin and bear it. No one likes a whiner. You are bound to learn something you don't know, and you also might meet some interesting and helpful people.

Get to Know Some Good COTRS

Other COTRs can be a good source of insight into the business. One of the best courses we ever took was a commercial one that reviewed the basics for COTRs from all sorts of government agencies. We got a bit of flack for it because it wasn't "Agency specific," but that was precisely why we wanted to take it. The class included COTRs from a wide variety of agencies such as the National Park Service and the Postal Service, and we were interested in how our problems compared with theirs. For example, what constraints were we under that they weren't? How did they solve problems? What drove them nuts in their bureaucracies? While the context was often different from ours, we gained many new insights from the experiences that these other COTRs described.

Our favorite story from that course was a Park Service COTR who found out that a contractor had buried his dog in the basement of an historic structure before pouring the basement slab. When the COTR confronted the contractor, the individual vehemently denied the claim. So the COTR told him to dig up the newly-cast concrete, and if there was no dog, he would personally pay to replace it. If the dog's carcass was there, the contractor would pay. Needless to say, the COTR didn't have to pay a penny out of his own pocket.

Get to Know Some Good Contracting Officers

There is nothing quite as effective as a like-minded CO-COTR team. Contracting officers are motivated by many different things, and you need to find those who are motivated by the same things that motivate you - like failing forward fast. When we started with the Agency, it was common for an office to have its own contracting staff that worked in the same location as the COTRs. This arrangement was effective because the familiarity fostered better teamwork, communications, and sense of purpose. Later, this system was replaced with a centralized, corporate-level contracting staff that was supposed to "improve efficiency." As you might guess, centralization did not work very well. We were stuck using a contracts staff that - quite frankly - didn't care much about us. Eventually we were able to craft effective operational arguments that allowed us to move our efforts away from these indifferent contracting officers.

One of the challenges we faced throughout our careers was that contracting officers rotated into and out of their assignments quickly. By the time we established a good working relationship with a contracting officer, that person was off to something new. The problem became much worse after the fall of the Berlin Wall when lots of experienced people left. New contracting officers were hired, but many of them didn't last. As a result, we were frequently dealing with inexperienced individuals who were more interested in making sure they never made a mistake than in getting things done.

The best strategy was to treat new contracting officers well and to try to develop a reputation for being careful and diligent. The payoff was that some of these inexperienced officers showed up later in our careers and we could work with them again on a good footing. Those prior experiences saved time, since trust and understanding were already established.

Procurements

Most contracting officers and managers will tell you that competitive procurements are the lifeblood of the Agency and that more is better. But the case is not all one-sided. Competitions are far from a sure thing when it comes to picking the right contractor, in part because writing proposals is not the same as doing research and development.

Contractors spend considerable time, effort, and resources on credible proposals. It is not fair to dump unnecessary proposal costs on them just because a COTR won't make the effort to winnow down the list of bidders, or a contracting officer is unwilling to back a good sole-source selection for fear that someone might object. Too often, the competitive process is required simply because it is viewed as the least controversial approach, especially when the low bidder is chosen.

If you must do a formal competitive procurement, make sure you do it right. The most difficult challenge in competitive procurements is to establish meaningful selection criteria. We have both been on many evaluation teams where (after all the results were in) several members said that the wrong contractor was selected because of the limitations of the evaluation criteria used in determining the winner. This simple item - making sure that the evaluation criteria match the needs of the program - almost never gets the attention it deserves. The criteria must ensure that marginal proposals do not slip into the competitive range. Our biggest personal complaint was that we were never allowed to use past performance on related efforts as a very significant selection criterion. In our experience, there is no better predictor of a future effort than a contractor's performance in a similar previous one. The rest of the world seems to understand this. When you need a new roof, for example, do you decide on a contractor based mainly on prior work or what some salesman tells you?

In the intelligence world, buying the management line that "competition is better" without thought of the consequences to the operation is irresponsible. Competitions increase security risks by exposing too many people to the problem and the technology. This, in turn, can make the transfer of technology to operations more difficult once the problem is solved. Tech transfer to operations was one of the biggest challenges we faced. If we blew that step, all of our time and effort was wasted. To minimize the transfer problems, we tried whenever possible to do the enabling research at companies the operations people already worked with. If the company lacked a particular capability, it could usually be subcontracted in some secure way.

If significant sensitivities are involved that require minimizing the number of people exposed to the problem, then push as hard as you can for a sole-source contract. As long as you do sufficient background analysis to more than justify your decision, then you have done the best you can for the organization and the operation. Even if you cannot convince management to allow the sole source, your information should help to reduce the number of potential bidders who are exposed to the details of the problem.

In our own work, our most successful projects were usually sole-source efforts done by small, flexible, and highly capable teams that remained largely intact over many contracting cycles. Most of these teams simply got better and better over time as they met new challenges. Security remained tight because there were few replacements, and transitions to operations became easier as the ops folks became more familiar and trusting of the technical teams.

The key to good sole sourcing and working with exceptional teams was avoiding complacency. Clearly, we needed to plan well enough to keep the teams funded. But even though we were usually satisfied with the performance we had, we constantly looked for new and better talent and technical capabilities, knowing that not all of these teams would remain at the top of their game. Just because a sole source is working doesn't mean you should put your research into alternative sources on hold. Also, don't fall into the trap of accepting mediocre performance just because you know the team. When you do find a promising new contractor, try them out on a small project that includes experimentation and prototyping. See how they do and how well they work with you. Don't ever bet the ranch on an unknown.

Resist the Urge to Copy Prior Work

Drafting good proposal requests (RFPs) can be hard, especially when you are new. There is great temptation to just mimic something that an earlier COTR did. While there is nothing wrong with reviewing the work of others to get an idea of what is needed and how things are typically worded, this should be tempered with sufficient discussion and study. You need to understand why certain things are in an RFP, where the numbers came from, and why certain wording is used. We have known many new COTRs who simply changed a few numbers in RFPs that were drafted by others without really understanding what they were doing. In effect, they accepted someone else's box as their own, with all its flaws and hidden assumptions. You need to rethink everything and not blindly copy someone else's work.

Understand the Unwritten and Unadvertised Rules

Never assume that you know all of the pertinent rules and regulations that govern the contracting process. In addition to the ones that your organization tries to drum into your head, other lesser-known rules also exist. Some of these rules are unwritten policies that happen to be in favor at the time, and others are contracting subtleties that you were never taught. Discovering these contracting insights is often just a matter of luck, such as being in the right place at the right time or asking the right question of the right person. Here are a couple of examples where the right questions helped us produce better contracts and results.

Firm-Fixed-Price Contracting

Although we started our R&D careers using the standard competitive procurements that led to cost-plus-fixed-fee contracts

(CPFF), eventually the sole-source, firm-fixed-price contract (FFP) became our mainstay for the same kinds of work. This might seem unusual for research and development contracts that have considerable risk, but it made sense in our world where unwritten laws of contracting also mattered.

The basis for our change in thinking was an unwritten rule that we discovered in a meeting with Agency lawyers. The lawyers were asked a question about appropriate contract types, and one of them explained that the contract type really didn't matter. From a legal perspective, all contracts were viewed as best effort. As long as the contractor demonstrated reasonable diligence in executing the contract, he would be paid and could not be forced to cover overrun costs - even on a firm-fixed-price contract. We had never heard this in any Agency training courses, and modified our strategies to take advantage of this new information because a firm-fixed-price contract was generally the easiest for contracting officers to implement.

New Work in Active Contracts

We discovered a second valuable unwritten rule in another of those legal meetings. This one involved the nasty problem of extra work done by the contractor that the COTR may or may not have authorized. If a contractor claimed that the CO or the COTR or any Agency manager gave verbal direction on executing work of any nature on an active contract, and the contractor followed what they were verbally told, Agency lawyers always sided with the contractor. The lawyer who explained this to us acknowledged that there were often disputes over whether the contractor was actually told that the new work was "authorized," usually with vague language that the work was: "a good idea," or it "needs to be done." The Agency policy of siding with the contractor in these disputes reflected an institutional view that our contractor base was essential to doing our mission and that it was not worth getting into numerous battles with contractors over what was said or not said. To minimize the possibility of any such misunderstandings in our own projects, we always insisted that contractors include a statement in their monthly reports citing that "no new work was authorized or done during this time period."

Slush Funds

Some years ago we attended a seminar hosted by the Directorate of Operations (DO). One of the speakers was the Deputy Chief of DO Contracts Staff, someone who dealt daily with contracting problems. He suggested that every contract should contain a task called trade studies (or some variation of it), and that this single task could represent up to 15% of the total contract value. He then explained that this was your "slush fund." It was the money that kept you from the delays in adding funds later to the effort for unforeseen problems. His view was that no contract is without problems, so plan accordingly. No one had ever explained this point in all of the COTR training we had gone through. There was a break after his talk and his associates descended on him like locusts, chastising him for telling us this unwritten rule. He replied that we needed to know this for all of us to be successful. Perhaps Contracts had to relinquish some control with this strategy, but it would also result in less work for them and more success overall.

Hybrid Contracts

A hybrid contract contains several types of contracts within the same vehicle. We were introduced to this form of contract when management ordered us to put in place a competitive contract for National Security Agency (NSA) work, using our contracting staff and our security envelope. NSA recommended four contractors for the competition, but provided no indication as to who would do the best work. Knowing the problems of full-blown competitions and betting too much on one solution early in the process, we went to the chief of our contracts staff for help.

The chief suggested a solution that nicely embodied a failing-forward-fast philosophy. He suggested a multi-year hybrid contract in which phase 1 (year 1) of the contract effort would be a firm-fixed-price demonstration of the technology and phase 2 (years 2 and out) would be a cost-plus-fixed-fee effort for the development of the system. Each of the contractors bid the first year and the out years as part of their original proposals, and they had the option of revising their out-year bids after the first year. This structure ultimately provided us with an efficient way to choose among the options. At the end of the first year, we received four demonstrations that provided good insight into the strengths and weaknesses of the different approaches. Two of the contractors were selected for the next phase and asked if they wanted to rebid or stand behind their original bid for the remainder of the contract. Of course, they both wanted to rebid with a 15% increase in funding to complete the work. We then offered them a choice. We could exercise the original option the next day or we could all wait six months until our contracts staff could negotiate a new contract. Not surprisingly, both contractors decided that starting immediately was more important than the possibility of increased funds. An immediate contract allowed them to keep their teams together, thus minimizing cash flow issues and improving their chances of a successful development.

In a world of unwritten rules and poorly advertised options, you must keep asking questions and looking for new and better methods. Don't assume that you have contracting figured out just because you have been doing it for a decade or more.

Options and Flexibility

Flexibility is vital to any failing-forward-fast activities, and one of our most effective contracting tools was the option task. Exercising options was a lot easier than starting a new contract, and we were at our failing-forward-fast best when we were mostly managing options. We constructed contracts with some number of options and used them to quickly change the direction of the work as needed. While inexperienced COTRs might have trouble with this method, it can save time, money, and many headaches. We typically defined a primary option for a contract that addressed the best solution known at the time and also a secondary option that had the same cost and schedule as the primary. Options for consulting work and risk-reduction studies were also added to help solve unanticipated problems without modifying the contract. Each option could be triggered multiple times. The key was that options could be quickly activated as needed without creating a new contract.

We recall an Agency customer who was attending a contractor review of a project to design and build a complicated system for his operational use. The customer asked how the effort would be done, since it required many new technologies and pieces working together. The contractor looked at us and asked if he should tell the customer the truth. We said "of course," and the contractor replied that we were going to make as many mistakes as fast as possible and determine the best direction to go. Our customer frowned for a second and then said, "Of course there is no other way to do it." Honesty prevailed. There was no contrived discussion of how smart we were or how well planned our schedule was - just the admission that we would adjust for the best result given the time and money that we had. In some ways, you can think of failing forward fast as reality contracting.

A Caution about Legal Approvals

As you have probably figured out by now, one of the things that the Agency did well in educating us was to create opportunities for discussion with senior people in the procurement chain. In another of these productive meetings a question was raised about what a signature from our legal office meant on a contract. The answer surprised a lot of the audience. The official explained that the sign-off meant only that the contract vehicle was legal but not necessarily that the tasks within the contract were also legal. Some or all of those tasks might require additional review and legal authorization before the work could be done. This was a pretty important distinction - particularly for an operational organization - but neither of us could recall ever hearing it properly discussed in COTR training. Legal is not necessarily common sense.

The Bottom Line on Contracting

We believe that without the authority and ability to manage your own contracts, you will likely never be able to implement a failing-forward-fast strategy. During our careers we saw many of the smartest technical whizzes flounder in their missions because they had to depend on other people to get contracting work done. You'll be more successful if you can gain as much control and influence over the procurement process as possible in your organization. If the government would have allowed us to qualify as contracting officers as well as COTRs, we would have signed up.

Recognize that while you will probably never be a contracting or legal expert, you need sufficient knowledge and judgment to ask the right questions at the right time. As our experiences suggest, keep asking questions of lawyers and contracting officers long after the mandatory courses are done. Those unwritten and poorly advertised rules that never show up in courses are often the key to the flexibility you need in a failing-forward-fast environment.

CHAPTER SEVEN

Some Things You Should Never Do

There are many ways to subvert the failing-forward-fast environment that you are trying to create. Here are some things to avoid.

We all have regrets about some of the things that we did in our careers. No one really escapes this burden entirely, but here are some thoughts that might help to minimize the pain and keep you failing forward fast.

Never Kid Yourself into Thinking That You Are Smarter than Everyone Else

Edward Snowden was a government contractor who leaked classified information about government surveillance programs in 2013. He was not the first person to violate his secrecy agreement and go directly to the press because of what he believed. In essence, he alone decided that his beliefs at the time were more important than peoples' lives, future intelligence from secret collection systems, large investments of taxpayer money, and the life's work of many employees in the Intelligence Community. Wow! He figured all of that out on his own. What an ego. He should have read our section on self-deception.

In the classified world, government employees and contractors sign a number of secrecy agreements, depending upon the work they do. In addition, civilian government employees swear or affirm that will "support and defend the Constitution of the United States against all enemies." Sometimes, you, as a government employee, might believe that your responsibilities conflict. But that is no reason to take matters into your own hands and cause damage that might or might not be worth the perceived wrong. The right phrase here is "beyond my pay grade." If something is bothering you, no matter whether it is theft, abuse, incompetence, or a belief in principle, do not try to get attention by leaking your work or that of others to unauthorized sources. This is one time when you really need to go through the proper channels, including the Inspector General (IG).[56] The classified information you have is not yours to leak.

Never Completely Trust the System to Keep a Secret

In any organization, you must be very careful about what you put out there that other people can see. In our time, we worried most about emails and other digital information because of the ease with which it could be shared with others without our permission or knowledge. Perhaps the most important lesson that we learned was to never respond to any message in anger or haste. Chances were really good that the next time we saw our message, it would

---

[56] Of course, you don't run to the IG for everything. We were once told by a highly-respected senior manager that he needed the IG's help four times in his long career.

be in an equally toxic reply that copied half of the world. If you must resort to a screaming match, do it in the privacy of your office. Government isn't reality television.

Regardless of the medium, you should never assume that things said in confidence will remain confidential. Live by the need-to-know principle.[57] Just don't be surprised if something you said in confidence is suddenly today's gossip. It is going to happen once in a while, but at least you will learn who can't keep their mouth shut.

Never Let Your Ego Get in the Way of a Proper Apology

When you screw up, apologize graciously. Contrary to what some people think, an apology is not a sign of weakness, but rather a sign of strength and confidence.

Never Trust Everything to Memory

For most of our careers, we carried 3x5 cards or small spiral notebooks in our shirt pockets (and no, we did not own pocket protectors). We used them for taking notes when we were having discussions with others and when we were just thinking about things. Some people like to show off by never taking notes, but that only really works for waiters looking for a good tip. Taking short quick notes when someone is talking is flattering to the speaker. The action says you think that something they said was important and not to be forgotten. Taking notes helps you to

---

[57] "Need to know" is a basic security principle that requires someone holding classified information to determine if someone else has the proper need to access that information.

respond accurately and bolsters your   reputation for being careful and thorough in your work.

Clearly, one cannot write everything down when dealing with classified information, so a good memory can really help.   We always believed that getting a good night's sleep before a day of meetings was a reasonable way to improve our retention of what was discussed during the day. If we were exhausted and not really paying attention, how could we possibly remember all that was said?   Later, we learned that was only half of the story. Researchers now tell us that our ability to remember events of the day depends on getting a good night's sleep afterward, since memories are stabilized while we sleep.[58] Working all day on the West Coast and catching a few winks on the red-eye back to DC, therefore, might not be the best strategy.

Never Say You Will Do Something That You Do Not Intend to Do

You want to be known for doing what you say. Otherwise, you are just another bureaucrat who only talks a good game.

Being busy is no excuse. Near the end of World War II, for example, General George Patton's army was fighting its way through Germany. As part of that effort, they liberated POWs who had been imprisoned at Stalag Luft III.[59] Clearly, these were busy and difficult times for General Patton. He explained to the POWs that his lines were stretched and that he could not get them out at the moment, but promised to get them help within 7 days. Seven

---

[58] Robert Stickgold, "Review Article Sleep-Dependent Memory Consolidation," *Nature* 437 (27 October 2005): 1272-1278.
[59] Stalag Luft III was the site of The Great Escape, a tunneling effort in which 76 Allied POWS escaped from the prison camp during the night of 24 March 1944. All but three of these prisoners were later caught and 50 of them were executed by the Gestapo under orders from Adolf Hitler.

days after he said this, the troop transports arrived. Patton had his faults, but happy talk wasn't one of them.

We once met with a DARPA program manager who had an interest in robots that could operate in surf zones. We were interested in the same problem and got together to compare notes. During the meeting, the program manager raised several questions that required some homework on our part, and we told him we would get back to him in a week. When we returned a week later with the information, he nearly fell out of his chair. He could only say: "This is the first time any of you guys ever promised something to me and delivered it on time!"

Meeting your commitments should be a no-brainer. Do what you promise when you promise, and you save others time and effort. This will help you enormously by gaining a reputation as someone who gets things done.

Never Work for Bad Managers

Studies done in the CIA during our tenure suggested that many employees quit because of problems with their bosses. That was sad because the employees had other options beyond leaving the organization or constantly enduring a bad boss.

When your boss leaves and the new one is appointed, always give the new boss a fair chance. Rely only on your own experiences and observations to make up your mind about whether you've got someone you can work with or someone you need to avoid. Try to ignore what others have said in the past about the individual. Rumors often lack context, are typically distorted, and are often outdated. People make mistakes, and sometimes they actually learn from them. What matters is how the boss works now.

When you finally decide that you work for a bad boss, however, don't put up with him or her any longer than you have to. Find another job or at least a way to get out of the line of fire.

For example, you might be able to survive a bad boss if you have a good supervisor between you and the problem boss. This creates the option to hide. Just as some folks work to get "face time" with the big boss, you can actively work to get less. We used this strategy to great advantage at various times during our careers.

Another option is to run. At one point in our careers, for example, our boss was going to be replaced with someone who was a problem. We had worked for the replacement before and all indications were that he had not changed. Since we did not want a repeat of past history, we decided that it would be a great time to return to the laboratory to update our lab skills and get some hands-on experience with new technologies. With a little foresight and planning, we were able to spend three wonderful years at a national lab and not return until the boss was replaced. With more than a thousand miles between the lab and the office in Northern Virginia, encounters with that bad boss were rare.

Never Be a Yes Man

Good managers count on their employees to give them honest feedback. While it can be easier in the moment to tell bosses what you think they want to hear, placating them is not the best solution when something really matters. When you disagree, make sure you do so respectfully, acknowledging the pros and cons of each position. If you can't put together a coherent argument on the spot, promise to return as soon as you think it through.

Of course, you also need to use some common sense and avoid criticizing everything your boss does that you disagree with. Some issues are more important than others. The most critical for you will likely be those that involve your projects. As a COTR, you are the one who is legally responsible for your contracts, not your boss. He or she can't change anything on the effort. The responsibility is yours to refuse when you have to.

The Agency had its share of insecure managers. Any management that only wants to hear their words parroted back to them isn't worth working for. Reasonable and effective managers want to know when they are making a serious mistake, even if it might not seem that way to you at the moment.

Never Be Abusive

Today, American society is a lot more sensitive to abuse than it was a few decades ago. We took our share of beatings at the Agency, both for things we deserved and things that we did not deserve. Sometimes, it got pretty nasty. We recall one instance where we had a meeting with our boss and his boss, a deputy director known for going off the deep end. On the way to the meeting, our boss remarked: "Promise me that you won't quit during this meeting." He knew what was coming and that it would not be fair. We took the abuse, but worked harder in the future to avoid that deputy director, since he seemed to enjoy beating up on people.

Being abused is degrading, but it does help you in the sense that it reveals a great deal about the person dishing it out. That insight can help you decide your next move. Most people go off the deep end once in a while, especially when work and worry get intense. So you shouldn't mind getting reamed out occasionally when you deserve it even a little bit. But when it is frequent, when it is disproportionate to the cause, and especially when the abuser seems to be enjoying the show, it is time to get out.

Never confuse a good knock-down and drag-out debate for abuse, however. We don't usually praise lawyers, but one thing that they are sometimes known for is how to argue with each other and still remain good colleagues. Technical people need to learn to do this as well. Be passionate about your position, defend it with everything you have, and - win or lose - leave it at the whiteboard (or whatever passes for a whiteboard in your future).

When emotions do take over, try to keep your mouth shut. It might feel good in the moment to tell your boss that the only time he opens his mouth is to change feet, but you are going to regret mouthing off later. How you handle difficult situations affects your credibility, which in turn affects opportunity and effectiveness.

It should go without saying that a COTR should not be abusive to contractors. Unfortunately, we've seen more than our share of COTRs who treated contractors like dirt. This never made any sense to us, particularly in a failing-forward-fast environment where you need a great deal of trust and honesty to make proper decisions.

Never Take Credit for Another Person's Work

In the late 90s, we were interviewed about someone who was a candidate for a very senior position within the Intelligence Community. We were working for the individual at the time, and the key question was whether this candidate was just in the room when things were happening or did he make things happen. This was a very wise question since many are perfectly willing to take credit for successes that they had little to do with, other than to show up. For the record, the candidate always did far more than just show up, and was always willing to give others credit. He was eventually hired, was a great boss, and we later spent the better part of 10 years working for him. Be honest about your contributions and expect others to be honest about them as well.

Bosses need to be careful that they give credit to their employees when it is due. Some bosses go overboard on this by providing too much public praise, but that is not nearly as bad as those who put themselves ahead of their employees. One such encounter occurred when we were staffed to the Deputy Director for Science and Technology (DDS&T). We worked for a boss who in turn reported to the staff chief for the office. At the time, we were struggling with a technical issue involving new computers that should never have been purchased. In one of those unexplainable bursts of insight, however, we figured out a way to replace them with better and more reliable systems at no net cost or embarrassment to us. So we explained the idea to our immediate boss. Less than 15 minutes later, the staff chief came running to us saying: "I've just heard the idea of the century," and proceeded to tell us our own idea. We realized very quickly that our sleaze-ball boss had just taken credit for our idea with no attribution. At that point, we started planning our way out of the staff job.

Why we didn't correct the staff chief? That would have clarified the issue, but it would also have made an enemy out of our immediate boss. Sometimes you just get screwed. Besides, things rarely happen in isolation in an office. Before telling the idea to our immediate boss, we had reviewed it with one of our colleagues who also worked for him. As a result, that colleague also got some valuable insight into the boss's character.

Never Hide Your Mistakes

People often get into more trouble concealing activities than confessing to them. Our best advice for when you have screwed up is (if time allows) figure out how to make sure it doesn't happen again and then tell your boss. It'll go a lot better if the boss sees that: (1) you know you screwed up and are taking responsibility for it, and (2) you are making changes to ensure that it never happens again.

Clearly, you should not try to place the blame on others. For example, we were handed an unclassified document to review late on one Friday evening. With the weekend coming up, we threw it in our knapsack to look at later. "Later" was Sunday morning at home when we discovered that someone had stapled a highly-classified document to the back of the unclassified document. Stapling two documents together is not unheard of, especially when people are busy or distracted. But the security violation was our fault, not the person who accidently put the documents together. We should have checked each page before assuming it was okay to take home. We handled this violation by going to Security, explaining what happened, and implementing a procedure that we would use to avoid any recurrences. It never happened again.

Never Be Your Own Lawyer

As we noted earlier, common sense is not always legal sense. You might need several meetings with lawyers to discover this, but you will come to believe it over time.

As a result, you should never make legal decisions that aren't confirmed by the appropriate lawyer. Get these decisions in writing because recalling a verbal discussion later on will seldom save you. Program managers who decided to make their own legal decisions often caused problems for themselves and the Agency. For example, a manager once told us that wartime powers in contracts allowed the government to take ideas from one contractor and give them to another. Since we were involved in Desert Storm at the time, he decided to use this power without asking anyone.[60] The truth was that he wanted to take someone else's good idea and give it to his favorite contractor. This was wrong and he is no longer managing contracts because of his actions.

And finally, don't let your boss be your lawyer. Don't assume that what he or she says is always okay. Verify the advice through proper legal channels.

Never Speak before Thinking

Warren Buffett once said that it takes 20 years to build a good reputation but it only takes 5 minutes to destroy it. You can make sure that you don't sink yourself by taking some time to think about what you are about to say.

---

[60] Desert Storm was a conflict with Iraq during George H. W. Bush's presidency. It was also known as the Gulf War.

CHAPTER EIGHT

Working Your Way out of a Mess

You can't plan for every contingency in a failing-forward-fast environment. Here are some thoughts about getting out of that hole you just dug.

No matter how careful you are, there will be times when you are truly stuck. Sometimes the issue will involve your internal work environment - perhaps a coworker is trying to sabotage you, or you have a serious disagreement with your boss. Other times, the issue involves the work itself. Perhaps your program is in financial trouble, your technical plan didn't quite work out, or your best engineer has decided to date a North Korean. While you do your best to prevent these problems from occurring, you cannot dodge them all.

You must learn how to fix these problems because the consequences go far beyond the immediate issue. They are the moments that define you as a COTR. Your handling of these issues determines to a large extent how your coworkers and contractors see you.

Don't Make Things Worse

You can make things worse in two basic ways, either by not doing something about an issue or trying to deal with it too quickly. The former is sometimes referred to as omission bias. This bias is toward doing nothing about an issue and letting things play out as they will. Individuals who work this way believe that not acting on an issue is somehow better than taking action. As Hutson notes in *The 7 Laws of Magical Thinking*, omission bias is a bit fatalistic. Let the gods decide.[61]

For others, it seems more natural to deal with a problem quickly because they are in pain and want the pain to stop. But one of the best ways to get deeper into trouble is to not think things through before acting. Don't accidently throw gasoline on a fire. Take the time you need to get it right.

Don't Blindly Follow the System

Always try to find the easiest legitimate way to solve your problem. Don't blindly accept some standard process or advice that someone gives you because that might not be the best way to accomplish your goal. Some things matter and some things do not, and be sure you know the difference. As we noted earlier, your career is not a test to see if you can follow a recipe.

In the late 1990s, for example, we struggled with a problem involving peer review of classified work. By its very nature, classified work is insular. Relatively few people know about the work and those who do tend to have some skin in the game already. While red teams, review panels, and more informal

---

[61] Matthew Hutson, *The 7 Laws of Magical Thinking* (New York: Hudson Street Press, 2012), 231-232.

checks and balances existed at the time to counteract possible error and bias, we believed that our officers would become better in their fields if they could experience the same sort of peer review that academics and others in the unclassified world relied upon.

One option we considered was to create a peer-reviewed scientific and technical journal for the Agency.[62] We ran the idea passed a number of folks that we trusted, honed it based on their comments, and then went to our office director. He liked the idea and told us that if we wanted to push it further, we should put together a PowerPoint briefing and start to march it through the office and up the management ranks in the S&T directorate. We left thinking that we had to find a more efficient way to get approval.

We eventually found our answer in a mid-level management course that we attended. The course was hosted by the DCI for the purpose of providing lower-level managers with a broader perspective of the Agency. The final exercise for the course had participants form teams and work with one of the four deputy directors on a management issue. We chose the Deputy Director of Science and Technology (DDS&T) team because we saw this as an opportunity to push the journal idea. Two important facts made this work: (1) no one on the teams really wanted to work that hard on the exercise, and (2) the deputy directors really wanted to take action on a suggestion if it was any good. These two forces effectively drove the process, so it was a short and easy effort for us to get the team to focus on peer review of work as an issue, prepare the pitch, and have it accepted. With the DDS&T behind it, everyone lined up behind the idea. We were able to implement our idea without an endless series of briefings, and the idea ultimately endured.

We are not suggesting that you disregard the rules. We are just saying that there are often several legitimate ways to get to the

---

[62] While this seems like an easy idea, it is not. Some of the challenges were how to balance need to know with sharing of information, how to incentivize officers to publish, and how to pay for the process and infrastructure.

same point. In our example, the other offices still got to weigh in on the journal idea, but it was from a top-down rather than a bottom-up perspective, which was much better from our viewpoint. There are useful opportunities everywhere, and it is just a matter of recognizing them and fitting them to your needs.

Seek Advice

One of the best ways to solve a problem is to talk with someone about it. Even if the person has no helpful advice in the end, the act of explaining out loud often provides clarity. For reasons we fail to understand, some people are very reluctant to ask for advice. Perhaps they see it as a sign of weakness or an admission of vulnerability. Or perhaps (as we have mentioned before) they view life like a test in college where you had to face the questions alone.

Once you start asking for help, however, don't accept everything you hear as gospel. Always consider the source. Try to cultivate a group of well-informed individuals in your virtual organization who know you and trust you and are known for their common sense. These are the folks you want to bounce things off in confidence. But don't mindlessly do what they say even if they have a long history of effective advice. You also need to think things through for yourself.

How do you tell good advice from bad? Here are a couple of rules that we have found to be helpful:

- Be wary of advice that is anecdotal. Our favorite red flag is "We tried that once and it didn't work." Such a comment can be useful if it provides you with a better understanding of the limitations of the idea. But you need to dig deeper. "Something" was tried, but sometimes it might not be exactly what you are thinking about and it was done 10 years ago before critical enabling technologies existed.

- Always assume that bias exists. Ask yourself whether the person providing the advice will benefit from a particular solution.
- Make sure that the advisor is not playing to you. Sometimes, people will tell you what they think you want to hear just to get rid of you. Your best and bravest advisors will tell you what they think, irrespective of what it might do to your relationship.

Yogi Berra once said that "You can observe a lot by watching."[63] Our version of this saying is that you can get a lot of advice just by watching. Look at your coworkers. Are their problems similar to your problems? What are they doing about them?

Go to Your Boss

Bosses can help you, and the really good ones enjoy doing it in a way that teaches you lasting lessons. Shortly after we began working in ORD, for example, we went to an Agency customer to talk about some of our work that could help them in the future. At the meeting the customer went ballistic (we sometimes said "nonlinear," but "crazy" also works). He said that he could not understand why we were doing this work, that he would call our boss to tell him that his office didn't support the work, and that it should stop. Keep in mind that his office was not paying for any of the work, had no obligation to act on it, and would get the results for free. Back in the office, we told our boss about the encounter. He just smiled and explained that the customer didn't like the idea because it wasn't their idea. They wanted us to stop

---

[63] Yogi Berra, *You Can Observe a Lot by Watching: What I've Learned about Teamwork from the Yankees and Life* (Hoboken, NJ: John Wiley and Sons, 2008).

so that they could have sole possession of the work. He told us to forget about the discussion and continue the work.

Our boss was right. People want to be successful and some will go to extreme lengths to make sure you don't compete with them. Sometimes they do this by the threat of influence and sometimes by deliberately making your work look bad. Not all bosses are as wise as ours was. For example, later in our careers we were told by a less-enlightened boss that: "The only reason you're here is to make me look good."

If you have the right kind of boss, there is nothing wrong with going to him or her with a problem. The trick is in how you do it.

First, don't bother your boss with every little thing. It isn't healthy to have a boss involved in every decision. We recall one DARPA director who was known for his desire to make every decision personally. This created one of the most dysfunctional times we can recall at that agency. By moving every decision up the chain, all lower-level interactions between program managers and customers and collaborators were stalled. Unfortunately, those lower-level interactions were an important part of DARPA's innovation.

Second, never go to your boss empty-handed. If you have a problem, make sure you understand it thoroughly, have asked the obvious questions, and have some options to suggest. Even if the options aren't great (you are stuck, after all), they show your boss that you have tried to work it out.

And finally, try to seek advice rather than direct help. This is your problem, not your boss's, so try very hard not to put the monkey on his or her back. A good line for ending a meeting is: "Thanks, this helps a lot. I can take it from here."

Argue with Your Boss

You cannot entirely avoid having arguments with your boss, and if you try to, you are not doing your job. For example, when your boss is about to do something really wrong, you have to step up. All arguments are not equal, however, and how you respond should depend on what the consequences are. For example, at one point in ORD, we saw a mid-level boss about to make a mistake involving one of the employees we were responsible for. The problem had racial overtones, and we pointed this out. He rejected our argument, and this created another problem. Was the issue settled at this point? Or did we need to keep trying? In the end, we had two more discussions, with increasingly refined arguments. Later, the boss came back and thanked us for being persistent. It was his way of saying that he had his head up his rear end and was really grateful that we helped pull it out. Our relationship was solid from then on and continued for years - all because of one bout of honesty and determination.

Other issues can take you closer to the mat with a boss. For example, a particularly nasty encounter for us was prompted by an upcoming television interview that our boss was preparing for. We had earlier shown him some video describing a technology that we were developing with another agency. It was stunning stuff, and he naturally wanted to use it for the interview. We refused because we did not have the authority or permissions we needed to release the video to the public. If the video wound up on television, it could impact relationships with universities, contractors, and the other agency that was providing some of the funding. That conflict led one of us to threaten to quit. Fortunately, the other of us mediated this disagreement and we eventually all got over it. You have to accept, however, that some of your interactions with bosses are going to be really intense when you disagree and the consequences really matter.

127

When Agency employees are promoted into the Senior Intelligence Service (SIS), they take additional training that is sometimes referred to as "charm school." When we took the course, we used it as a means to better understand how the promotion system really worked. SIS is a big deal and getting promoted into its ranks is hard, so what better audience to answer promotion questions? As part of our survey, we asked 20 of our classmates if in the past two years they had a knock-down discussion with their boss sufficiently intense that they thought they would have to leave or be told to leave. How many of them said that this was the case? Surprisingly, all of them![64] It is not about getting along all of the time. Good management wants the truth. Of course, there are good ways and bad ways to tell the truth, and the outcome can depend on how tactful you are.

Accept Criticism Graciously

We all know people who do not take criticism well. How many times, for example, have you seen people give talks and then react defensively to the comments and questions? Later in our careers we spent a fair amount of our time critiquing research work done by relatively inexperienced COTRs. The level of comment and suggestion that we provided ultimately depended on the individual's reaction to it. If the COTR took the suggestions and came back for more, we would bend over backwards to help them. If they tried to blow us off or discount what we advised, we gladly let them flounder. Darwin is at work, even in the government.

Criticism is important to a COTR, and we think that one of the finest compliments is that someone has a thick skin. Criticism can be unfair and personal sometimes, but you need to let it roll off of

---

[64] This led us to believe that when some Agency employees reached the GS-15 level and stayed there for years, they quit holding back on their bosses rather than worrying about getting promoted. At that point, they became more promotable to SIS.

your back. You want to encourage free and open discussion about ideas and problems, and this will only occur if you are grateful for the advice and consider it carefully. Perhaps the key is to understand that - in most cases - people are criticizing your solutions, not you personally.

Recently, we drafted a report that was going to be given to a number of individuals for review. At the meeting, the document was revised to the point where it was very different from where we started, and some team members were apologetic. They felt that we had put a lot of work into the original and thought we would be offended by the changes. To us, we only saw a process making something better, and we were very happy about it. The tragedy would have been if the others had not engaged and just accepted the draft as final.

Try Some of These Tricks to Get Unstuck

Your background, knowledge, and history - and the biases they engender - all contribute to your view of a problem. These help to form the box in which you are trapped, and getting out of that box often means that you have to deal with those biases. Unfortunately, there is no sure-fire recipe for altering perspective and coming up with good solutions, but there are some tools that can help you. The following are a few that we have used with some success.

Approach a Problem as Someone Else

When faced with a problem, we always preferred to develop ideas together because this kept each of us from falling back into our own comfort zones. When you are on your own, however, you need to pull yourself out of that zone and provide perspectives that are missing. One way to do this is to first list the ideas that emerge from your background, and then ask the question: what would someone else have come up with? Essentially, you force yourself to role play another person, thinking about his or her areas of expertise and what solutions they might create. If you are more comfortable viewing things from an electrical viewpoint, for example, then look at them from a mechanical engineer's perspective. This might sound a bit strange at first, but it is no different than the times in your personal life when you faced a problem and asked: "What would Mom or Dad do?" If you want to see past a problem, approach it as you think someone whom you know well would.

Use Analogies and Linkages

Analogies and linkages are ways to relate something that we already know to something that is new. Scientists and engineers have understood the power of analogy for many centuries. Albert Einstein, for example, relied heavily on analogy to help him with his thinking on special and general relativity. Some, like Douglas Hofstadter and Emmanuel Sander, believe that analogy is central to all thinking.[65]

In some cases, it helps to link a particular problem in one field to a different field. For example, scientists and engineers have for many decades produced new products based upon biological structures and processes. Velcro (inspired by the burrs that collected on a dog's fur) and riblets (drag reduction surfaces inspired by shark skin) are two such examples. By understanding why natural materials and systems have particular structures and properties, scientists and engineers can take advantage of ideas that nature took millions of years to produce. Such biomimicry has been of benefit to scientists and engineers in fields as diverse as robotics, aeronautics, computer science, and environmental engineering.

---

[65] Douglas Hofstadter, and Emmanuel Sander, *Surfaces and Essences: Analogy as the Fuel and Fire of Thinking* (New York: Basic Books, 2013), 3-32.

In other cases, just linking your problem to a new technology could help. In the 1970s, for example, the maturation of digital imaging technology completely changed satellite reconnaissance from film-based collection systems to real-time electro-optical systems. Among other advantages, this new technology eliminated the delay caused by the need to recover film capsules and process film.[66] For a problem today in 2015, for example, can a contribution from nanotechnology, or bioengineering, or 3-dimensional printing, or crowd sourcing help to solve your problem?

Try Humor

Humor can help to change the direction of your thinking. Humor is often an exaggeration of reality, and exaggeration can sometimes give you a new insight into whatever you are doing. It is also relaxing and can help to ease any tensions that might be forming in a group. Don't think of jokes as a distraction. Good answers to really tough problems are rarely reached by going in a straight line.

Try Puzzles

While doing puzzles might not help you with your immediate problem, working them as a hobby can help you develop skills for dealing with the next problem. Doing crossword puzzles, for example, teaches you the consequences of bad assumptions and how to find and remove them - such as the plausible but wrong answer to 24 across. When you are stuck in the real world it can sometimes be due to incorrect data or assumptions. Requirements and crossword puzzles, for example, have a lot in common.

Puzzles are very effective in demonstrating that each of us unconsciously builds mental barriers or boxes when solving a problem. We've all experienced that "aha" moment when we finally find the right viewpoint and the solution to a puzzle

---

[66] National Reconnaissance Office, *The National Reconnaissance Office at 50 Years: A Brief History,* http://www.nro.gov/history/csnr/programs/NRO_Brief _History.pdf (accessed April 15, 2015).

becomes clear. Remember the classic puzzle that challenged you to connect the dots in a 9-dot square array using four straight lines?[67] The key constraint was to not lift the pencil or retrace a path. Most people try unsuccessfully until they realize that the lines can go beyond the border formed by the dots. You can probably blame that particular bias on all of that time spent coloring within the lines in kindergarten.

Try the Other Way Around

R.V. Jones suggested an elegant trick for seeing problems differently in a paper entitled "The Other Way Around."[68] His advice was that when you have a design for something, try inverting it in order to gain additional insight. During WWII, for example, Great Britain developed a wide range of escape and evasion tools for use by prisoners of war in Europe. Accurate concealable maps and compasses were essential escape aids. The compasses were concealed in hollow buttons that unscrewed to provide access to the device. These special buttons were sent to POWs inside of aid packages, along with other clothing and food. Unfortunately, the Germans caught on to this ruse and began inspecting buttons coming into the camp. By this point you might be able to figure out how the British got around this dilemma. They simply replaced the right-handed threads with left-handed threads on the buttons.[69]

Try Jones' trick next time you are in a technical review. We did this during a briefing about a robot whose mobility was based on a new type of drive mechanism. As we examined the device, we subconsciously recalled the other-way-around idea. When we described the inverted design to the engineers, they were surprised

---

[67] Wikipedia, *Thinking Outside the Box*, https://en.wikipedia.org/wiki/Thinking_outside_the_box (accessed September 20, 2015).
[68] R. V. Jones, *Instruments and Experiences: Papers on Measurement and Instrument Design*, (Chichester: Wiley, 1988), 460-474.
[69] Clayton Hutton, *Official Secret,* (New York: Crown, 1961),145.

that they had never considered it. They stopped thinking about alternatives when they found the current design.

Pose the Right Questions

Sometimes you get stuck because you are focused on the wrong question. At one point in our careers, for example, we were asked by an Intelligence Community research organization to develop an R&D plan for ground mobile robotics technology. This was shortly after NASA had successfully operated Sojourner on Mars, and folks were assuming that we should make more use of such devices as well. What should we be investing in? The organization was expecting an answer like communication links, location and mapping technologies, mobility methods, etc. We did a vertical dive into the problem and surfaced a bit nervous. We saw many commercial and government funded robotic efforts and began to wonder. The core of our concern was the problem of not knowing how effective these devices really were, and particularly their specific weaknesses and limitations. We sensed that some of the contractors and program managers were in a bit of denial regarding the real problems with these devices. When we were shown a robot, for example, it would invariably get stuck, and the operator would release it from its self-inflicted prison. Such flaws hardly registered with the developers. Yet a robot that could not move competently was not of much operational use.

It was time to ask the question a bit differently. Did it really make sense to invest in the development of robotic technologies now? Sojourner was an amazing feat for its time, but NASA missions are not intelligence missions and Mars is not the Earth.

The questions that ultimately moved us and the committee forward were far different from the one we started with. They included the following. How do we learn the limitations of the technologies and pass those concerns on to the developers? How do we teach our operations personnel about robots and their limitations? How do we know when a robotics-related technology is ripe for use or for adapting to our specific problems?

Using these questions, we discarded the idea of a comprehensive development program in robotics technology and systems, and proposed a different effort. Our alternative proposal was for a testing facility that provided two services: (1) mission-relevant standard tests of any existing robot or robot technology, and (2) the ability to cobble together surrogate missions for those in the community so that they could get a better feel for whether a robot-assisted mission made any sense. Both of these services encouraged a failing-forward-fast philosophy for those involved in robotics of interest to us and later led to specific developments. The facility successfully performed its mission for about 15 years, and enjoyed a number of cooperative agreements with other agencies.

Of course, it is sometimes hard to know what the right question or problem is. The quote by G.K. Chesterton that: "It isn't that they can't see the solution. It is that they can't see the problem" is true more often than we would like.[70] We were lucky with the robots.

The most important general question you can ask yourself when you are stuck is whether you are solving the right problem. There is no point in struggling with a problem if solving it leaves you on an incorrect pathway. It is easy to get started on the wrong track with wrong requirements, vague committee questions, or the wrong contractor, but it is also possible to have a good plan go bad due to outside trends. In our careers, the largest of these game-changers occurred when the Soviet Union collapsed. The Soviets had been our primary adversary since the Agency was created in 1947, and much of our research was targeted against them. If there was ever a time to ask if we were solving the right problem, that was it! And many of us did, abandoning research areas that were no longer warranted, nursing along areas that we were uncertain of,

---

[70] G. K. Chesterton, *The Scandal of Father Brown* (New York: Dodd, Mead & Co., 1935), Chapter 7.

and emphasizing areas that applied to a diverse set of other targets such as terrorism. Not surprisingly, however, a number of our colleagues had great difficulty with this question and continued to try to defend work that was no longer necessary. If you work in a failing-forward-fast mode, you understand that it is counterproductive to hang on to things that no longer really matter, just because you are comfortable with them or have much of your life invested in them.

CHAPTER NINE

Staying Healthy in a Tough Job

Keep your nose clean and one day you'll be an office director like me eating dinner out of a vending machine...Bob H.

It is easy to neglect yourself when you are doing things you believe in. Before you know it, you are no longer the "new guy," and in what seems like an incredibly short period of time, you become the "old guy who has been around here forever." Everyone gets the signal some point. For us, it was a cafeteria in La Jolla, California. After going through the line with our selections for dinner, we went to the cash register to pay. When it was our turn, the young lady operating the register smiled at us and asked if we were eligible for the senior discount. At that point we knew what it meant to have time stop, as we stared at her and thought: "Well, it had to happen sometime."

Time Passes Quickly and Damage Is Cumulative

When you're young, you seldom think about your health. You recover from a cold quickly and the flu keeps you down for a week at most. But your recuperative powers lessen over time, and as you get older, you generally take longer to mend. Trying to stay healthy can mitigate this inevitable decline, and starting early can also help.

The key to taking better care of yourself is recognizing that your work does affect your behavior, your eating habits, and your sleeping habits. Do you, for example, eat very fast or do you enjoy leisurely meals with lots of conversation and perhaps a good bottle of wine? At work, do you eat at your desk or snack excessively instead of getting away for a while? Do you consume a lot of coffee or soda instead of drinking water? Do you eat a lot of sweets and other vending machine food? Do you take antacids or more powerful reflux drugs for heartburn? Do you have trouble sleeping? Many engineers and scientists will probably answer "yes" to more than one of these questions. We recall one offsite where some Agency scientists and field operations people had dinner together, and it was readily apparent who was a scientist and who was a field person. The scientists had finished eating within 15 minutes and were excusing themselves so they could use their cell phones while the field personnel were still leisurely eating and enjoying the company of their colleagues.

We have both learned the hard way that you have to invest time and effort in staying healthy. Don't give up. Sometimes there is nothing you can do about a particular affliction, but in most cases you can do something that will make the circumstances better. Although the literature on exercise and healthy eating is quite large and complex and often contradictory, we think that most of you can probably get by with just a bit of common sense and some willpower. While we have no particular scientific expertise in wellness, here is what worked for us:

- Don't sit for a long time. Being a COTR means sitting on your rear end in front of a computer or in meetings for much of the day. Credible research shows that all sorts of chemical changes start to occur when you plop your butt in a chair. Some folks, including former Secretary of Defense Donald Rumsfeld, stand when they do desk work. We don't stand, but we use other tricks. For example, we keep our printer in the basement and our computer system on the first floor. Getting a printout involves getting up and taking a short hike.

- Get your heart working a bit by taking a 30-minute walk every day, or riding a bike, or doing calisthenics.

- Get some high-quality sleep because it is essential to your basic mental and physical health. In addition, it reduces the likelihood that you will nod off and start snoring and drooling in the middle of a meeting - always a bad career move.

- Keep muscles toned and bones strong. This does not mean that you need to join a gym. Your body is a perfectly good set of weights. Stretching exercises are good, and so is mowing the lawn with a mower that does not propel itself.

- Try to eat healthy food, but don't become a fanatic. For example, Mountain Dew and a powdered sugar donut are always a bad breakfast choice, but whether you eat only organic foods or not probably doesn't matter very much in comparison. One thing that helped us later in life was cooking our own food from scratch rather than purchasing something from a restaurant or buying prepackaged meals in a grocery store. There are clear nutritional benefits to cooking food from scratch, but the psychological benefits (including taking your mind off of work) might be the real payoff.

- Don't ignore the possibility that some foods can cause you problems and yet be fine for other people. If you suspect some of your problems are coming from certain foods, experiment on yourself and find out.
- Don't neglect your brain. The good news here is that what is generally good for heart health is good for brain health.
- Have some friends whom you can rely on.

If you are married or in some other committed relationship, work hard to keep it alive. Doing classified work can be tough on home life.

24-Hour Accessibility

Being a COTR is a 24-hour-a-day job. You always have issues to deal with and they are always in the front or the back of your mind. The problem is exacerbated, however, by today's communications where anyone with a problem can try to reach you at any time of day or night. We are a far cry from the days when no one had your home phone number and work was done from 9 am to 5 pm.

Interruptions always have a larger impact than you think. Someone calls with an issue, there's a conversation, and then the call ends. But it doesn't stop there. Interruptions have time constants, so it takes time for the effect to die away. If the subject is a simple thing like confirming a meeting, the effect dies quickly and you can get back to whatever you were engaged in before. On the other hand, if the subject is a security issue, it might gnaw at you for hours, or maybe the whole night.

We do not have a great solution to this problem of 24-hour accessibility. In clear emergencies, you want the call. But some folks have lousy calibration points when it comes to what requires an off-hours discussion. Try to be explicit with them about what constitutes an emergency and perhaps resort to using the blocking modes on your smart phone. You also might want to reconsider working with folks who can't be recalibrated.

Taking Vacations and Working Weekends

When each of us was nearing retirement, we both had maxed out on leave that we could accumulate. In retrospect, that was stupid. While you want to have a buffer for extra vacation and family emergencies, you need to take most of the vacation days that you are owed.

It is important to have a real life outside of your work. You should not put off the things you like or want to do until retirement. If you love to fish or camp or hike, for example, take some long weekends or other blocks of time each year do them. Don't think that you will work hard and "catch up" on the fun when you retire. Anything that involves strength and physical wellness is at risk with this strategy. At 65, you might be able to hike around the lake, but hitting all of Colorado's fourteeners might be out of the question.

We think that a bucket list is a good idea, but you should create it early in life and modify it as you age. Creating one will quickly help you to identify those things that are best done young, and those that are more age independent. One of the big benefits of having a bucket list when you are young is you will find that many of the things you just "had to do" didn't turn out to be as great as you thought and you can quickly move on and replace them with something else. In some ways, this is simply an application of the failing-forward-fast approach to your life.

It almost goes without saying that if you are spending most weekends at the office or are using most weekends for business travel, then you are doing something wrong. Some weekend work is unavoidable, but you shouldn't routinely sacrifice your family, your friends, or your life for your job. Weekends are your best chance (short of vacation) to get work out of your head for a while. Everyone needs occasional breaks, so don't be tempted to use most of your weekends to catch up on office work.

Your physical and mental health should be your most important project.

CHAPTER TEN

Some Thoughts about Retirement

If you are on the fence about retirement, the one thing we can tell you is that the water is fine. Come on in.

We have several colleagues who will likely leave this world at their desks. That is their preference, although we think it is a bit sad that work is so much of their lives. We both prefer to die somewhere else.

Retirement can be a grand adventure if you plan for it properly. Both of us went through the process of deciding to retire from government and having spent a few years as retirees, now have some thoughts on this matter. If you don't want to die at your desk, our perspective might help.

Making the Decision to Leave

People start to think seriously about retirement for many reasons ranging from health, to opportunity, to "this place is driving me crazy." In our own cases, changes in management and the world situation focused us on retirement. We were convinced in the end by simply making a list of the senior managers that we would be willing to work for. When that list shrank to one, we knew it was time to leave.

For an increasing number of government employees, however, personal health issues or the declining health of a loved one will drive the decision to retire. While the government is relatively generous when it comes to family medical issues, health concerns can eventually dominate your life. When that happens, it can be tough to do a good job at work. A leave of absence is a solution in some cases, but probably not for illnesses that can span many years.

Saving Money for Retirement

There was a time when government provided a really good retirement. You did not pay into Social Security, but instead gave the Civil Service Retirement System (CSRS) 7-8% of your pay. For this investment, you received an inflation-adjusted pension worth roughly 2% of your highest-three earning years times the number of years you worked.[71] If you worked for 30 years, for example, you received a pension of over 55% of your high-three salary. It was clearly a good investment.

---

[71] The actual formula provides 1.5% for the first 5 years, 1.75% for the second 5 years, and 2% for all years after the first 10.

Ronald Reagan changed all of that when he instituted a new retirement system known as the Federal Employees Retirement System, or FERS. This put federal employees into the Social Security System, shrunk the pension to roughly half, and gave employees the opportunity to participate in a form of 401(k) program known as the Thrift Savings Plan (TSP). The new system was designed to provide about the same level of benefit as the CSRS system, but required much higher contribution rates.

In the future, FERS is likely to change for the worse, as government finds ways to reduce future costs. Even if you put the maximum in throughout your career, therefore, you should assume that it will not be enough. Remember that this is the same government that created the traditional IRA as an incentive for workers to save for retirement, and then did not change the maximum allowed contribution for over two decades.

There is no substitute for retirement saving early in life. While it is hard to predict what the economic situation will be when you retire many years in the future, you can only decide based on what you know now. Sure, you will likely get hurt by politicians at some point, but alternatives such as doing nothing or saving just when you feel like it probably won't work well at all. At the very least, the federal retirement system provides a structure in which to save a portion of your income.

Finally, don't ignore how things get taxed. The TSP and traditional IRA are good in the sense that they defer taxes and allow you to grow money with those deferred funds. They are bad in the sense that those contributions and earnings (which include capital gains) are, with some exceptions, taxed at your regular income tax rate at the time of withdrawal. Roth IRAs are also a good alternative tax strategy for many earners.

Whether you decide to save money inside of the federal system or outside of it, or both, just make sure that you start young and keep after it. Nothing conveys freedom like a buck in the bank.

New Life, New Rules

A new set of rules applies in retirement. The old bosses and the bureaucracy are gone, and they are replaced with a different set of influences. At this point in your life, you probably have more control than ever before, and you shouldn't squander that opportunity. In our own cases, one thing we did was to write this book. We did it for a couple of reasons. For one, we thought we could provide a perspective that was a bit different from what others have written about the Agency. The second reason was that it helped us to reflect a bit on many of the things that we had developed and thought about over the years. Retirement gave us the quiet time that was always in short supply when we were working for the Agency. Time provides clarity. It also helps you let go of the grief and trauma that you might have experienced. We can now laugh a bit about all those bad managers and goof-ups that inevitably happened.

We have known many people who feared retirement. A common story in the Agency was about someone who retired and was dead within a year. That does happen, of course, but mostly it does not. Retirement can be the best years of your life. We know one person who has been retired for a very long time. In her view: "I've outlived 3 husbands and 8 dogs and these are the happiest years of my life." Her thing is teaching bridge to the elderly, and she never thought that it was possible to be so happy in retirement.

Pay Attention to What Happens in Washington

After 25 or more years in government, the last thing you probably want to do is pay attention to what Congress is up to. Unfortunately, you need to keep an eye on what is happening to your retirement and healthcare benefits. Don't assume that you are going to continue to get paid what you are owed just for having a pulse.

At some point, you should consider joining the National Active and Retired Federal Employees Association (NARFE). They work aggressively to protect your benefits and let you know when something that is going on in Congress could affect your retirement. They are also very good about answering specific retirement questions.

Do Not Retire to Nothing

Retirement is a slap on the face in several ways. First, your ego will take a hit, and second, you will (perhaps for the first time in decades) find yourself with time on your hands.

Retirement is a very effective experiment for finding out who really likes you and who really likes what you were able to give them when you were in government. Depending on how good you were at assessing people, this will either be a surprise or a confirmation. While rejection can be very hard on your ego, it is actually good because it is both humbling and therapeutic. You now have a chance to prove yourself in new ways.

Try to avoid doing more of the same old/same old. Don't become a consultant to your old place unless they really need the help. Work for someone else, try volunteering, learn to cook, and build that boat you've been thinking about for years.

There is no reason to stop thinking about failing-forward-fast just because you are retired. Use the time to experiment with new things, to fail, and to learn more about yourself. Our experiments so far have included: learning how to play bridge (and meeting a very different group of friends in the process), running a farm (and gaining a whole new appreciation for rural life), volunteering in community activities (and learning that volunteer work is very hard) and participating in local politics (and learning that the squabbles at the local level can be just as vicious as at the federal level). We both rejected the idea of second careers as sumo wrestlers, however.

Loose Lips in Retirement

People who worked in classified environments have some special issues to worry about in retirement. Your guard will go down when you are no longer involved in day-to-day classified work. In addition, your ego might start to take a hit since you are no longer in the know and in the thick of things. Yet people will often ask you what you did for a living and there will be a temptation to go further than you should. But you can't. It is dangerous to think that the information is old and that no one cares. What harm could there be in telling a story or two that makes you look good? But you don't really know. So behave yourself.

Be Kind to Those Who Will Outlive You

While all of us want to die quickly and painlessly when our time comes, the reality suggests that many of us are in for a significant period of time when we can no longer be independent and will need help with daily living. Use some of those planning skills you developed in the government to anticipate what you will need as you become less independent.

Everyone needs a durable power of attorney for medical care, a living will, a durable power of attorney for finances, and a will or trust:

- A durable power of attorney for medical care identifies someone you trust to make medical decisions for you when you can no longer do so for yourself.
- A living will specifies what measures will be taken to keep you alive when you are terminally ill.
- A durable power of attorney for finances identifies someone you trust to make a variety of legal and financial decisions for you when you can no longer do so yourself.
- Wills and trusts specify how your assets will be dispersed after your death. Done well, they can greatly ease the work of your executor.

You don't necessarily need a lawyer or tax accountant to help you with these documents, but you do need to spend time thinking deeply about what you want and understanding the limitations and consequences of these documents. Create them while you are fully competent. Do not wait until you are ill.

All of this probably sounds a bit depressing, but you need to remember that death and disease and frailty are normal parts of life, and accept that that they are going to happen. Ignoring reality does not make them go away.

A Memorandum for the Future Director of CIA

MEMORANDUM FOR: Future Director of the Central Intelligence Agency

FROM: Bruce and Greg

SUBJECT: Our Take on the Path Forward

It is impossible for us to pass up an opportunity to provide some caution about the future to those running the organization. We believe that significant change is needed, and here are some of our thoughts.

- Your organization is too slow for the missions you are running. Figure out how to extend a failing-forward-fast mentality to all parts of the Agency and its contractor base. Your people must not fear failure if they are to adequately meet the future intelligence needs of the United States.

- Force your middle and lower management to deal with the problem children. Too many times, upper management knows someone is a problem, but doesn't do anything about it until things hit the fan. Once you deal with these problem people, you can start to trust the remaining people with more responsibility.

- Without sustained research and development, you will lose much of your edge in understanding technology and using it effectively. It might be that the old ways of doing research no longer apply as well as they used to, and change is warranted. The important first question, however, is not how to organize research and development in the Agency, but rather what the functions and products of R&D should be in the future.

- The IG needs to be closely monitored because it is usually an employee's last resort.

- When someone asks you or your managers for funds to harvest 'low-hanging fruit" remember that the fruit had to come from somewhere. If you don't plant the seeds and water them, there is no tree and there is no chance of fruit.

- You need to go "back to black" in much of what the Agency does, how its people think, and how its tools are used. Tradecraft needs to be better protected. In the future, you will increasingly rely on many of the same tools that the rest of the world has, and your advantage will largely come from how you use them.

- Ignore the temptation to reduce the number of people with clearances because they are an integral part of a failing-forward-fast strategy. Instead, do a better job of managing need to know.

- Get better at flexible planning. Resist the bureaucratic tendency to treat any strategic plan as a box check. Also be suspicious of any planning document that is just an extension and justification of current efforts. Planning is a messy and uncertain business.

- Do a better job with training by putting deception at its core. As a former Soviet case officer once told us: "isn't it all about deception?"

- Run the Agency with the assumption that you have moles all of the time and make sure employees act cautiously as well.

- Stop the "everyone gets a trophy" mentality with medals, because this dilutes the respect that the real heroes deserve. People who risk their lives for their country deserve medals. Bureaucrats who do their job deserve bonuses and corner offices.

- Figure out how you are going to run an Agency with fewer people and more technology, just like the rest of the world.

- When someone in your organization does something spectacular, ask yourself whether it happened because of the organization or in spite of it. And then act accordingly.

Good Luck

About the Authors

Bruce Hartmann and Greg Moore are 25-year veterans of the Central Intelligence Agency where they spent their careers developing devices and concepts for use in Agency operations.

They can be reached at:

KJI Ltd
PO Box 9257
Pueblo, CO 81008